Doing Good Science
in Middle School

A Practical STEM Guide

Including 10 New & Updated Activities

Expanded **2nd** *Edition*

Doing Good Science
in Middle School

A Practical STEM Guide

Including 10 New & Updated Activities

Expanded 2nd Edition

Olaf Jorgenson
Rick Vanosdall
Vicki Massey
Jackie Cleveland

NSTA press
National Science Teachers Association
Arlington, Virginia

National Science Teachers Association

Claire Reinburg, Director
Wendy Rubin, Managing Editor
Andrew Cooke, Senior Editor
Amanda O'Brien, Associate Editor
Amy America, Book Acquisitions Coordinator

ART AND DESIGN
Will Thomas Jr., Director
Rashad Muhammad, Graphic Designer

PRINTING AND PRODUCTION
Catherine Lorrain, Director

NATIONAL SCIENCE TEACHERS ASSOCIATION
David L. Evans, Executive Director
David Beacom, Publisher

1840 Wilson Blvd., Arlington, VA 22201
www.nsta.org/store
For customer service inquiries, please call 800-277-5300.

FSC
www.fsc.org
MIX
Paper from
responsible sources
FSC® C011935

NSTA is committed to publishing material that promotes the best in inquiry-based science education. However, conditions of actual use may vary, and the safety procedures and practices described in this book are intended to serve only as a guide. Additional precautionary measures may be required. NSTA and the authors do not warrant or represent that the procedures and practices in this book meet any safety code or standard of federal, state, or local regulations. NSTA and the authors disclaim any liability for personal injury or damage to property arising out of or relating to the use of this book, including any of the recommendations, instructions, or materials contained therein.

PERMISSIONS

Book purchasers may photocopy, print, or e-mail up to five copies of an NSTA book chapter for personal use only; this does not include display or promotional use. Elementary, middle, and high school teachers may reproduce forms, sample documents, and single NSTA book chapters needed for classroom or noncommercial, professional-development use only. E-book buyers may download files to multiple personal devices but are prohibited from posting the files to third-party servers or websites, or from passing files to non-buyers. For additional permission to photocopy or use material electronically from this NSTA Press book, please contact the Copyright Clearance Center (CCC) (*www.copyright.com*; 978-750-8400). Please access *www.nsta.org/permissions* for further information about NSTA's rights and permissions policies.

Library of Congress Cataloging-in-Publication Data

Jorgenson, Olaf.

  Doing good science in middle school : a practical STEM guide / Olaf Jorgenson, Rick Vanosdall, Vicki Massey, Jackie Cleveland. -- Expanded Second edition.

      pages cm

  Includes index.

    ISBN 978-1-938946-07-3—ISBN 978-1-938946-63-9 (e-book) 1. Science—Study and teaching (Middle school) 2. Technology—Study and teaching (Middle school) 3. Engineering—Study and teaching (Middle School) 4. Mathematics—Study and teaching (Middle School) 5. Inquiry-based learning. I. Title.

  Q181.J69 2014

    507.1'2—dc23   2013042891

Cataloging-in-Publication Data for the e-book are available from the Library of Congress.

Contents

Contents

Acknowledgments

The authors wish to thank author, leader, and educational innovator Doug Llewellyn for his wisdom, encouragement, resourcefulness, and guidance throughout our comprehensive rewrite of this book. Thanks also to our peer editors and master teachers, Jacqueline Kimzey and Bruce Jones, to our technical advisor John Hunt, and to our families for their patient support.

Dedication

For Juliette, Wesley, Grant, and children everywhere who benefit from good science and great teachers; and to Kash, Krew, and Paige, who are part of the next generation of science learners.

About the Authors

OLAF (OLE) JORGENSON served as director of K–12 science, social sciences, and world languages in the Mesa Unified School District in Mesa, Arizona. Previously he was a teacher and administrator in U.S. and international schools, mostly at the middle school level. Ole is past president of the Association of Science Materials Centers (ASMC) and served on faculty with the National Science Resources Center's Leadership Assistance for Science Education Reform (LASER) strategic planning institute, with a focus on middle school science issues. He has presented on middle school science reform at the National Science Teachers Association's annual meeting and at ASMC's Next Steps Institute, and is a past member of NSTA's National Committee for Science Supervision and Leadership. Ole's other publications focus on topics in instructional leadership. He holds a doctorate in educational leadership from Arizona State University. Ole lives with his wife, Tanya, and their daughter, Juliette in San José, California, where he is head of school at Almaden Country School. He can be reached via e-mail at *olafjorgenson@yahoo.com*.

RICK VANOSDALL is a professor in the College of Education at Middle Tennessee State University and also serves as interim director of the doctoral program in Assessment, Learning, and School Improvement. Rick has held various teaching and leadership positions in preK–12 and higher education across the past three decades, including principal investigator on various education service and research projects funded by the National Science Foundation, the Department of Health and Human Services, and the U.S. Department of Education. Rick has delivered national and international conference presentations and is active in science education action research. Rick earned his doctorate in educational leadership at Arizona State University. He lives in Tennessee with his wife Kim, son Grant, and daughter Wesley, along with the family pooch, Sedona. He can be reached via e-mail at *rvanosdall@gmail.com*.

VICKI MASSEY completed her 35-year teaching career in the Mesa Public Schools by serving as the district's secondary science specialist. Vicki currently teaches science methods courses at Northern Arizona University (NAU) and Arizona State University and does science consulting and professional development around the United States. She has served as past president and board member for the Arizona Science

About the Authors

Teachers Association, director of NSTA Division XIV and NSELA Region D, and board member for the Arizona Science and Engineering Fair. Vicki received her Masters of Arts in Science Teaching from NAU. Her experience as a curriculum developer and professional-development specialist deepened her appreciation for science teachers and the tremendously important job they do each day. Vicki lives in Mesa, Arizona, with her husband, horse, and ever-loyal pug and border collie dogs. She can be reached by e-mail at *vickimassey@cox.net*.

JACKIE CLEVELAND completed her career in education as a K–6 science specialist in the Mesa Public Schools, where she also served as a basic skills specialist. She is the recipient of the Presidential Award for Excellence in Science Teaching and a 1991 semifinalist for Arizona Teacher of the Year. During her tenure, Jackie held positions as the National Science Teachers Association preschool/elementary director and president of the Arizona Science Teachers Association. Jackie taught science methods courses at Arizona State University, Northern Arizona University, and University of Phoenix. Jackie lives in Gilbert, Arizona, with her husband Neal. She can be reached via e-mail at *cleveland1@cox.net*.

Background

This book grew out of the authors' experiences while they worked in the Mesa Public Schools (MPS) in Mesa, Arizona. MPS is a metropolitan district serving 70,000 students and 90 schools, K–12. Mesa has used learner-centered methods, inquiry principles, and hands-on science units since 1974, with a science resource center for kit development and distribution starting in 1979. Mesa's science program has earned praise from the National Science Teachers Association, *Harvard Educational Review, Newsweek, American School Board Journal, Parenting* magazine, *American Scientist,* and *The Executive Educator*. In the past two decades, the district has celebrated five awardees of the Presidential Award for Excellence in Science Teaching.

The district's science program and resource center was developed by longtime director Dr. Susan Sprague, now a semiretired science consultant living in northern Arizona. Mesa's resource center refurbishes and distributes more than 10,000 kits annually to its 55 elementary schools. The middle school program also includes two self-contained fifth-grade flight centers with aircraft and helicopter simulators and night-vision goggle stations, serving all of the district's approximately 4,000 fifth graders each year. To find out more, please visit the MPS Science and Social Sciences Resource Center (SSRC) website at *www.mpsaz.org/ssrc*.

Introduction
Changes, Changes, Changes!

Talk about an explosive decade for teachers and learners! When we wrote the first edition of this book, YouTube and Facebook had not preoccupied America's waking hours, and "Google" was not yet a verb. As teachers across our nation recognize, the rapid proliferation of digital technology and concurrent accelerated "flattening" globalization of the world's economy have reshaped the potential and purpose of American education in profound ways.

Middle school teachers today serve a new generation of children weaned on social media that didn't exist in 2004. They instantly access information in and out of the classroom that was previously hoarded, protected, and strategically dispensed by their teachers. It is, to say the least, a very different classroom world today compared to 10 years ago, when we wrote the first *Doing Good Science*.

And these seismic shifts—combined with the emergence of distance and blended learning, flipped instruction, interactive digital textbooks, the furious rise of standardized testing, the popularity and widespread embrace of robotics, development of the Common Core Initiative and the *Next Generation Science Standards* (*NGSS*)—all called us to revisit our work and evaluate how we might make the book more useful in the context of 21st-century skills that teachers today are expected to cultivate in their students.

We realized quickly that improving the second edition would require more than updating and freshening our original book. We've added chapters and sections to help teachers assimilate STEM principles, understand and apply the *NGSS* and *A Framework for K–12 Science Education* (including the new expectations for engineering design), and integrate the *Common Core* literacy and writing standards. We've also learned from teacher feedback about some holes in our first version, such as advice about how to arrange and conduct collaborative table groups and about the critical role of scientific argumentation that we've addressed in this new book.

Most important, since teacher-friendly, ready-to-use STEM activities are the core of this second edition, we've gathered more teacher feedback to provide 10 new and updated investigations aligned with the revised standards and reflecting the emergent emphases on engineering design, STEM, and the 5E method that were present but less pronounced in the original 10 activities.

Based on reader reviews from the first edition, we've also sought to *deepen* the activities as well—beyond explaining what each activity is about, we aim in the new set of activities to help teachers engender productive conversations with students about the investigations and their resulting data.

Throughout our work in rewriting the book, though, the same basic goals that motivated us to create *Doing Good Science* have guided our efforts in this version:

1. Provide a useful resource for science teachers of varying ability and experience levels, staying focused on colleagues in their first years of teaching middle school science, who serve students in a wide range of school and community settings and with differing levels of resources and support.

2. Keep the book readable and user-friendly, addressing the demands teachers face concerning integration of STEM and literacy skills.

3. Provide activities that address the standards, engage students (good science is *fun!*), and are neither costly nor dependent on access to science kits or expensive technology.

Above all, we want *Doing Good Science in Middle School* to be a book by teachers, for teachers, that can make what you do for your students a little easier, richer, and more enjoyable.

We hope you agree that we achieved our objectives. Enjoy!

—Ole, Rick, Vicki, and Jackie

Preface

A middle school science classroom was once described to us as "a nuclear reaction about to happen, on an hourly basis." At the time, that description was meant to illustrate the unstable, unpredictable, and at times irrational behavior of a group of middle schoolers. Years later, we know that the behavior in question is pretty typical, but it can be significantly more challenging to deal with when middle-grade students are confined to neat rows of desks and numbed by textbooks, teacher-centered instruction, and lack of meaningful interaction with peers or their teachers.

In this book, we propose opportunities for learning and teaching amidst the sound and fury of a different sort of explosive (but productive) middle school science classroom. In our experience, good science—by which we mean activity-based STEM instruction—promotes the unexpected and delightful development of adolescent middle school students.

For us, good science constitutes a shift away from the textbook-centered direct instruction that emphasizes discrete factual knowledge claims and passive observation of science phenomena toward active, learner-centered, hands-on and minds-on investigations conducted to some degree by students themselves. Good science and middle school learners are very compatible, as we'll explain in Chapter 1.

Who are we? We are four educators who worked together in Mesa, Arizona, in a school district that has embraced good science instruction since 1974. We are among those who have come to enjoy the blossoming intellects, often comical behaviors, and insatiable curiosity of middle schoolers and who *choose* to work with them! With more than 130 years' combined experience in the profession, we've gathered a lot of ideas to share. We know from our interactions with educators around the country that relatively few quality resources exist to assist science teachers "in the middle," and this was a central impetus for writing and then updating *Doing Good Science in Middle School*.

Our book is aligned with *A Framework for K–12 Science Education* (2012) and the *Next Generation Science Standards* (2013), which set forth eight practices that are fundamental to understanding the nature of science:

- Asking questions (for science) and defining problems (for engineering)
- Developing and using models
- Planning and carrying out investigations

- Analyzing and interpreting data
- Using mathematics and computational thinking
- Constructing explanations (for science) and designing solutions (for engineering)
- Engaging in argument from evidence
- Obtaining, evaluating, and communicating information

We've used the *Framework* (2012), *NGSS* (2013), and the *Common Core State Standards* (2010) as the basis for recommendations to assist middle-grade science teachers, while unpacking the *NGSS* to make them more easily accessible. Throughout the book, we kept in mind teachers who work in self-contained team formats as well as departmentalized middle school configurations.

Our work here is meant to meet other important objectives but above all, we intend it to be teacher-friendly. We wrote *Doing Good Science* as practitioners, for practitioners. In this book, you will find

- a comprehensive overview of science and engineering practices, STEM, and inquiry-based middle school science instruction, aligned with the *Framework* and the *NGSS*;

- a conscious connection to the *Common Core* literacy and math skills embedded in the *NGSS* that help determine—and are fostered by—student success in good science instruction;

- 10 teacher-tested activities that integrate STEM with literacy skill-building (with emphasis on safety in the science classroom);

- information on best instructional practices including argumentation and formative assessment, along with useful print and Web-based resources, science associations, workshops, and vendors;

- a solid foothold for new teachers to help them teach science and engineering practices while better understanding their often enigmatic middle-grade students; and

- an opportunity for veteran teachers to reaffirm that what they do is "good science."

We hope readers will find this book easy to use. It can be read in its entirety or perused section by section as a reference for lesson and unit planning and as a basis for evaluating and modifying existing lessons. It will help teachers explain to their principals why their classes at times need to be noisy, bustling, and "social" to be effective.

We also hope this book is for some readers a point of departure from relying solely on teacher-centered methods with passive text- and worksheet-dependent curricula and in favor of the active learning potential and rich teaching opportunities that good science makes possible in the middle grades.

Let the journey begin!

References

NGSS Lead States. 2013. *Next Generation Science Standards: For states, by states*. Washington, DC: National Academies Press. *www.nextgenscience.org/next-generation-science-standards*

National Governors Association Center for Best Practices and Council of Chief State School Officers (NGAC and CCSSO). 2010. *Common core state standards*. Washington, DC: NGAC and CCSSO.

National Research Council (NRC). 2012. *A framework for K 12 science education: Practices, crosscutting concepts, and core ideas*. Washington, DC: National Academies Press.

CHAPTER 1

Good Science and the Middle School Learner

Socialization, Autonomy, and Structure

W hether you are a veteran of the middle-level classroom or a relative new-comer, your success with students in these grades will depend on your ability to adjust instruction to the cognitive, emotional, developmental, social, and psychological demands of the middle school learner.

Even though their immersion in a digital world has further clipped their already scant attention spans, middle schoolers remain wired for the more active approach provided by good science instruction. As we stated in the preface, by *good science* we mean a shift from doing textbook- and teacher-centered instruction to conducting learner-centered investigations, framing and solving problems, engaging in field studies, developing and refining solutions for engineering problems, and performing experiments in which the teacher becomes a guide and resource rather than a director and source.

Science and Engineering Practices, as the *NGSS* term them, are the skills students need to develop to be able to "do" science and engineering. The practices are the mecha-nisms for varying degrees of student-directed discovery, autonomy, and heightened engagement in science and engineering activities. Using these skills, students acquire discrete content (called Disciplinary Core Ideas in the *NGSS*) and make connections across science disciplines via the Crosscutting Concepts.

Good science involves activity, choice, and independence. It also requires careful planning and structure that middle schoolers need and which allows them to thrive; we introduce these three main elements of the *NGSS* here because they need to be part of all that we do when planning good science lessons.

The Match Between Middle Schoolers and Good Science

Good science is highly compatible with the fundamental needs of adolescent learners, especially when their teachers understand the profound developmental changes with which these youngsters contend.

Conversely, traditional teaching methods that rely on textbooks, direct instruction, seatwork, PowerPoint slides, and lecture are successful depending on students' abilities to endure extended periods of concentration, inactivity, and careful note taking—none of which is a particular strength of most 10- to 14-year-olds we've known. However, such teacher- and text-centered instructional strategies prevail in the vast majority of hundreds of science classrooms we've observed.

Consider Table 1.1, which illustrates the compatibility we've seen between the developmental traits of adolescents and the characteristics of good science as we define it.

Active teaching methods are compatible with the way middle-level students naturally learn, and support their intrinsic tendencies in developmentally appropriate ways. Middle school youngsters are not well-served by a curriculum founded on passive lecture and worksheets. Awareness of the developmental levels and cognitive parameters of 10- to 14-year-olds is paramount to effective middle school teaching in any subject.

Table I.I. Compatibility Between Middle Schoolers and Good Science

Adolescent Traits	Good Science Characteristics
Curiosity and interest in learning	Develops questioning skills; emphasis on "doing" science
Varying cognitive levels (shift from concrete to abstract reasoning)	Uses multiple scientific practices, cultivates different learning styles
Need for relevance, to connect learning to prior knowledge	Emphasizes experiment, experience, problem-based learning
Increased sense of independence	Emphasizes discovery learning
Need for social interdependence	Emphasizes collaboration and cooperative, project-based learning
Relatively short attention spans; easily bored, easily distracted	Uses activity-based instructional design; rote learning de-emphasized
Need for validation; insecurity, fear of failure; developing self-concept	Provides opportunities for noncompetitive authentic assessments; fosters environment supportive of risk-taking
Simultaneous need for autonomy and structure	Offers degrees of teacher influence/guidance and differentiation
Need to be acknowledged as an "adult"	Learner-centered, rather than teacher-centered

Typically Atypical

Until the mid- to late-1990s, relatively little research focused on middle-level learners compared with studies of older and younger children. Indeed, there is such a range of maturity and development in middle school students that it is difficult to find agreement about what these young people should even be called: Adolescents? Emerging adolescents? "Tweens"?

Whatever we call them, middle school students are not children, and they are not adults; yet, they can display the delightful and aggravating characteristics of both groups, and in dizzying succession, with little or no way to predict which will happen next. One day they want recess, and the next they want to drive your car. Needless to say, while abundant research in developmental psychology happily explains their behavior profile, middle schoolers make life interesting for their teachers.

In his now-classic guide to happy coexistence with middle schoolers, Rick Wormeli observes that "young adolescents are moving through one of the most dynamic stages of development in their lives. As teachers, we might have to bushwhack through the hormonal tendrils on a daily basis, but it's worth the effort to find the gold inside each child" (2001, p. 7).

Adolescents are coming to terms with rapidly increasing knowledge of themselves, their world, and their place in it; in any group of tweens, maturity levels swing dramatically between individuals. Some are already launching into puberty and toward sexual maturity and may find it difficult to relate with their peers who still enjoy playing with dolls and Legos. Tweens are excited and excitable, though their engagement plummets toward the tail end of their short attention spans. Mood swings are dramatic and common; most middle schoolers are developing a more complex sense of humor; reactions are typically extreme and often accompanied by telltale teen eye-rolling, exasperated sigh-grunts, and the classic added syllable, "-uh"—as in, "NO-uh," "beCAUSE-uh," and more.

All of this is exacerbated by the impact of the digital media competing for their attention, interest, and hours that we'd like them to be sleeping.

For some youngsters, the middle school years are exciting and affirming, and for others, awkward and unsettling. Teachers at this level need to be especially sensitive and vigilant toward their students who manifest symptoms of depression or worse.

In any case, this sweeping range of social, physical, emotional, and intellectual maturation makes it difficult to generalize about "what works" in teaching middle school students, beyond the need to diagnose and monitor developmental progress throughout the school year. "Getting young adolescents to pay attention and learn is 80% of our battle in middle schools. The rest is pedagogy" (Wormeli 2001, p. 7). On a very basic level, teachers of middle schoolers should have a tolerance for ambiguity and a fundamental willingness to be flexible. (They also need, and middle schoolers really appreciate, a sense of humor.) To some extent, we believe teachers can learn to be spontaneous—to find comedy in inopportune displays of bodily functions, to lose precious class time to an evasive cockroach on the ceiling—but if a prospective middle grades teacher finds that her need to finish a chapter or get through a lesson overrides her willingness to tolerate the odd "accidental" belch or to spend a few minutes calmly swatting an overhead roach amidst much fanfare, she might not be cut out to work with this breed of student. Effective middle school teachers enjoy interacting with students at this age.

If flexibility and spontaneity are assets in a middle school teaching assignment, it is equally important for us to turn our attention to the adolescent need for structure.

Need for Structure

Given the profound, rapid, and sometimes confusing array of changes middle schoolers experience between grades 5 and 8, it is not surprising that they seek stability to varying extents and in different ways. Above all, middle-level students need the structure provided by classroom procedures because it makes school a safe, protected, predictable environment in a surrounding world that for many of them is as unstable and unpredictable as their own mood swings. Their newfound freedom is alternately thrilling and frightening, and their limited experience with autonomy leaves most adolescents in need of some degree of order and security. It is common for adolescents to fluctuate between demanding independence and welcoming direction from adults who they know care about them as individuals.

A substantial amount of our foundational moral development takes place during these years, when young people begin to associate actions with consequences and the broader ethical structures that support moral judgments. What is right, what is wrong, and why? How far is too far, and why? Thus, middle grades students depend on and will necessarily test the structure provided by classroom expectations and rules. If a middle grades educator successfully teaches procedures, students adopt the routines as their own so that structure (and rules) becomes a source of comfort rather than confrontation.

If, on the other hand, the teacher posts the rules without taking time to teach them, points to them when a student gets out of hand, and demands explanations for "misbehavior," there's bound to be continued testing and experimenting from the students ("accidental" belches and tipped chairs are favorites, for example). Middle schoolers understand the difference between control and support, manipulation and respect; they resent being treated like children, even when they behave childishly. Mutual respect is part of their worldview—and nothing cultivates respect more successfully than teachers who frame their expectations and procedures in terms perceived by middle schoolers as "adult to adult." In our experience, respect is most likely to be enjoyed by teachers who understand and set their expectations with attention to the fears, insecurities, goals, demands, and motivations of middle grades youngsters. (See more on classroom management in Chapter 5.)

Middle School Thinker: Not an Oxymoron

Most people don't typically perceive middle school-aged youngsters as "thinkers," at least not in the academic sense. Middle schoolers are by nature eager to seek and discover—but how they pursue and what they do with information they encounter can vary dramatically from one student to the next. Adolescent learners are scattered on the continuum between what psychologist Jean Piaget termed *concrete* and *formal operations*. Students still functioning at the concrete level have difficulty understanding the relationship between variables, for example. Those who have progressed developmentally toward formal operations can make inferences and begin to reason deductively. At that stage, students can design controlled experiments and determine relationships between multiple variables. Formal operational thinking leads to understanding increasingly complex modes of organizational schemata, such as the periodic table or the structure of DNA.

It is common to have a classroom with a few students who can reason only at concrete levels sitting next to students whose sophisticated, abstract thinking can make it interesting for the teacher to stay ahead from lesson to lesson. A one-size-fits-all, text-based approach to science does not readily accommodate the wide developmental differences between students in the middle.

We need to acknowledge that middle schoolers, regardless of their place on the cognitive continuum, ought to be treated as "thinkers" for two important reasons. First, the nature of science and engineering practices allows for—indeed, demands—that teachers design activities with a range of abilities and aptitudes in mind. Whether the student's role is to accomplish limited, concrete tasks as a member of a group conducting a complex project or to be responsible for a synthesis of a group's findings involving assumptions and flaws in the group's experimental design, inquiry activities are equally appropriate. Our sample STEM activities illustrate this point.

Second, middle schoolers should be treated as thinkers because no matter what level the teacher finds students to be functioning at cognitively, socially, and behaviorally in the first weeks of school, the teacher's task is to move them forward throughout the school year. We realize that this happens amidst much sound and fury during the middle years; middle schoolers are often distracted (and sometimes overwhelmed) by their emotional brains, making it extremely difficult for meaningful cognition to take place. There are many challenges to prolonged student attention in a middle school classroom—beyond sneaking peeks at smuggled digital devices—especially if accompanied by inactivity or inconsistent procedures. But again, as is the case with many aspects of effective middle grades instruction, regaining and maintaining student engagement mostly comes down to four good science habits:

1. Teaching clear procedures
2. Reinforcing routines
3. Providing opportunities for active learning
4. Developing and maintaining relationships

Passion for Discovery

Whether middle-grade science students are stalled by high school-level assignments that are still too abstract, or bored by elementary-caliber seatwork that is not at all challenging, we've observed that science is often made to seem hard, disinteresting, and irrelevant. Science in the middle grades needs to be fun, fundamental, and connected to the lives of adolescents. We've found that when we fail to meet their needs in this way, far too many youngsters in the middle grades are turned off to science.

When many students fail in the middle grades, whether in a certain class or in a sport or in their socialization, the impact is formative and, at times, life-altering. After all, it is in these crucial years that a child's dispositions toward learning are formed. (Am I good at math? Do I like to read?) In many students, negative attitudes toward science are bred in the middle years, can fester in the face of multiple failures, and will become progressively resistant to remedy in later, more difficult, and increasingly abstract science classes: "Research has shown that if educators don't capture students' interest and enthusiasm in science by grade seven, students may never find their way back to science" (NSTA 2003). Consequently, teachers in the middle grades are charged with lighting the fires of curiosity and cultivating the innate adolescent passion for discovery, rather than snuffing it out with too much lecture or too many worksheets.

We also know that 21st-century skills such as problem solving, teamwork, and analysis and evaluation of information hang in the balance during the middle school years. Good science can play an essential role by enabling students to practice and empowering them with successful experiences—because good science embodies 21st-century skills.

Good science in middle school calls for active teacher instruction. After the necessary planning and setup, a key role for the teacher is to move from group to group or lab station to station, helping facilitate investigations and experiments. Teaching good science is also intensive, no doubt about it, requiring more time for planning and preparation than traditional teaching. Good science demands patience, tenacity, passion, and the desire to grow as a teacher. But we've found that it requires comparatively much more effort to deal with boredom, behavior problems, and the *Ferris Bueller* effect ("Anyone? Can anyone tell me the answer? Anyone? Anyone?") that accompanied our early attempts to teach from a textbook. Indeed, once you have your students hooked on good science and familiar with the procedures, they do most of the work, and the teacher gets to enjoy the hum (well, roar) of excited young people *doing science*.

References

National Science Teachers Association (NSTA). 2003. *Science education for middle level students*. NSTA Position Statement adopted in February 2013. Arlington, VA: NSTA. *www. nsta.org/about/positions/middlelevel.aspx*

Wormeli, R. 2001. *Meet me in the middle: Becoming an accomplished middle-level teacher*. Portland, ME: Stenhouse.

Yager, R., ed. 2012. *Exemplary science for building interest in STEM careers*. Arlington, VA: NSTA Press.

Discovering the Nature of Science

Questioning, Argumentation, and Collaboration

The more one learns about doing good science in light of the traits of middle school learners, the more evident it becomes that middle schoolers are ideal candidates for activity-based investigations. Middle schoolers need activity, and they get excited about solving actual problems in teams—collaborating— as opposed to routinely completing worksheets in the stifling isolation of their desks. And it is all this *doing*, we contend, that is the prerequisite to discovering the nature of science.

In this chapter, we examine good science principles in greater depth with an emphasis on how they can be applied by teachers in classrooms and how collaborative science activities driven by questioning and argumentation can be productive with highly social, easily distracted middle schoolers.

Helping Students Grasp the Nature of Science

Ultimately, good science provides an experiential foundation on which scientific principles, terminology, and historical context, *stick*. As Jonathan Gerlach (2012) puts it, "We need to give students the opportunity to experience science before we start explaining science" (p. 47).

In our experience, students begin to understand the nature of science by *doing* science: observing, questioning, hypothesizing, determining patterns and cause/effect relationships, developing models, designing investigations, analyzing data, and constructing evidence-based explanations and conclusions.

In the process of doing good science and thinking like scientists, teachers lead students to ask and attempt to answer questions and frame problems accurately so they can devise models to solve them. These sorts of active investigations cultivate student curiosity, and we rejoice (quietly) when the kids who are "too cool for school" get hooked by one of our investigations or problems. Doing good science makes the *study* of science fun and relevant, and relevance is crucial to making science engaging for middle schoolers.

By conducting hands-on, minds-on investigations themselves, or with degrees of teacher guidance, students gain scientific experience that creates a context for appreciating the discoveries "real" scientists have made throughout history: the Copernican Revolution, Darwin's survival of the fittest, Marie Curie's radioactivity, George Washington Carver's numerous agricultural breakthroughs, Watson and Crick's structure of DNA, and more. When middle schoolers wonder and puzzle and discover, they make connections between their own experience and the thinking processes that led to these landmark innovations; thus, they gain insight into the nature of science in ways no textbook or lecture can provide.

In our experience, good science "works" with students of all ages and ability levels because of the match between its methods and the way we learn. Learning lasts when new information is connected to previous knowledge and experiences. Especially for middle grades learners, the experiential platform built by activity-based science provides relevance and context for the terminology, processes, concepts, and case studies from the history of science presented in text-based and teacher-centered instructional methods.

It is important that good science not be misinterpreted as something so prescriptive as "the scientific method"—as if there were only *one* method that scientists follow in their investigations. Good science activities call on students to acquire and apply scientific knowledge, to develop and use higher-order reasoning skills, and to communicate information and conclusions—but not by following any single procedure. These skills are applicable in any subject and grade level and cultivated by the careful questioning and problem-solving that characterize science and engineering practices, but they are much broader and more complex than a singular scientific method. Students who practice thinking like scientists discover the important roles that creativity and entrepreneurial exploration play in the discovery process, transcending and often departing from a step-by-step methodology. Such structured guidance may be more appropriate for children in elementary grades to lay a foundation for organizing scientific investigations.

Indeed, good science in middle school is defined by varying degrees of independent and creative thinking, as students learn to lead their own investigations and the teacher assumes the role of co-investigator. Their autonomy rests on *choice*, which Doug Llewellyn points to as a critical tool teachers employ as "an authentic means to increase intrinsic motivation" (2013, p. 90). "Effective teachers," Llewellyn asserts, "focus on fostering intrinsic motivation that encourages self-determined and self-fulfilled autonomous students, moving instruction from 'you have to' to 'you can choose to' and shifting the ownership of learning from the teacher to the individual student" (p. 90).

Those of us who've spent time with middle schoolers know how desperately they crave responsibility and choice; their fundamental, developmentally appropriate need and desire for autonomy shouts out the compatibility between good science methodology and middle school students. This is powerful stuff.

Asking Questions, Questioning Answers

We present strategies teachers can use to implement effective questioning in Chapter 6. Here, we look at the thinking behind the questions, and why effective questioning is so essential with middle schoolers doing good science.

The *Framework for K–12 Science Education, Next Generation Science Standards*, and *Common Core State Standards, ELA* all call for teachers to develop students' critical and argumentative abilities, encouraging them to pursue evidence-based arguments and challenge assumptions much as the original Greek skeptics did. In this way, teachers model questioning and argumentation, essential to the process of "doing" science, and enable students to develop the necessary science process skills that link observations, claims, and evidence. (One teacher's slogan—plastered on every wall in his classroom, under the clock, on the trash cans—put the message this way to his students: *ASK QUESTIONS. QUESTION ANSWERS!*)

What this teacher was doing, of course, was urging his students to step away from the hand-holding, passivity, and teacher-centric approach to learning that characterizes most classrooms in our country. Inviting middle schoolers to challenge authority? Is he nuts? Well, not if there are well-defined procedures in place to keep student energies (and irreverence) productively channeled. (There were!)

"To scaffold students toward more independence, teachers need to move from a classroom culture of control to a classroom culture of choice; from a culture of conformity to a culture of creativity; and from a culture of teacher dependency to a culture of student autonomy" (Llewellyn 2013, p. 95). Truly, "letting go" is harder for some of us than for others, as we move across the spectrum of guided- to less-guided investigations. It requires a big shift for some teachers, we realize.

In good science, the teacher directs safe activity without dominating and assists without answering. Asking reciprocal questions (what our students called "never giving a straight answer") drives students crazy at first; eventually they get used to it, and finally, they like it. But the reason we need to let go is because in the pursuit of professional science, no external authority provides hints to or reassures scientists, and the veracity of a scientist's conclusion, more often than not, is confirmed by her peers rather than someone outside the project design. In this way, science and engineering practices can mirror the experience of "real" scientists. You can't fool the students. Once they get the hang of questioning and scientific reasoning, and savor the confidence and independence that good science affords, they know the difference. Drill and kill is history.

Questioning, Inquiry–Style

Those of you who've been teaching for a decade or more may be surprised at the shift in emphasis in the *NGSS* away from inquiry. It's as if the term fell out of the *NGSS*, after we all heralded inquiry as the core of good science since the first science units emerged in districts including Highline, Anchorage, Fairfax County, and Mesa a half century ago. The *NGSS*, and the majority of the second edition of this book, reference science and engineering practices instead.

Harold Pratt, a renowned science education leader, offers the following explanation behind the shift in terminology:

Why do these practices replace the word *inquiry*? The change is described as an improvement in three ways:

- It minimizes the tendency to reduce scientific practice to a single set of procedures
- By emphasizing the plural practices, it avoids the mistaken idea that there is one scientific method
- It provides a clearer definition of the elements of inquiry than previously offered. (2012, p. 13)

While inquiry may be less prominent in the *NGSS*, the importance of questioning remains at the foundation of doing good science, accompanied in engineering investigations by careful definition of the problem(s) to be examined and addressed. For the purposes of our book, we hold that good science and engineering practices depend on good questions and clearly defined problems. (We look at how to use "focus" questions for scientific investigations and frame problems for engineering design activities in Chapters 4 and 6.)

Furthermore, we enthusiastically agree with Harold Pratt that engineering design and the other practices constitute a powerful context within which inquiry is one of many ways real scientists apply their work. The *NGSS* and *Framework* call for integrating science and engineering practices, crosscutting concepts, and disciplinary core ideas; we believe this integration affords teachers and students a fuller, multidimensional representation of good science compared with the model we presented in the first edition of this book.

We believe another reason it's prudent that "inquiry" is not specifically emphasized in *NGSS* and the *Framework* is that the term has lost meaning, hijacked over the past decade by many in and outside of K–12 science education who used it merely as a marketing device. That said, we wholeheartedly advocate for the authentic use of constructivist pedagogical practices, including inquiry. Additionally, we believe constructivist practices are essential for preparing students to meet the NGSS.

This is why, even though it is no longer the anchor theme as it was in our 2004 edition of *Doing Good Science*, we didn't completely eliminate inquiry from this substantial revision of our book. You will see the term—and the concepts—interspersed throughout the chapters.

Table 2.I. Big Shifts in Science Education

Less Emphasis on	More Emphasis on
Scientific process skills	Science and engineering practices
Science taught in isolation	Science as the crossroads of curriculum
One scientific method	Multiple methods of exploration
Knowing scientific concepts	Using scientific knowledge to solve problems
Science labs done in isolation	Problem- or place-based learning
Using text from hard-bound books	Accessing digital media
Summative assessment	Formative assessment
What student knows indicates level of mastery	What student can do indicates level of mastery

Armed with the full complement of science and engineering practices, crosscutting concepts, and disciplinary core ideas, moving a more traditional teacher, school, or school system toward activity-based science at any level still involves fundamentally rethinking priorities and principles. It also invites teachers to reconsider their roles and relationships with students as "co-investigators"—together sharing an investigation for which they do not necessarily know the outcome. (Even teachers may not be able to predict an experiment's results when students are charged to a greater or lesser degree with its design and implementation or with the task of designing solutions to engineering problems.) Generally speaking, good science lessons involve a shift in teacher responsibilities to the students when compared with traditional science instruction, as illustrated by Doug Llwellyn's (2007) glimpse in Table 2.2 at how inquiry lessons facilitate this shift.

Table 2.2. Invitation to Inquiry Grid

	Teacher-Initiated		Student-Initiated	
	Demonstrations	Activities	Inquiry	Inquiry
Posing the Question	Teacher	Teacher	Teacher	Student
Planning the Procedure	Teacher	Teacher	Student	Student
Formulating the Results	Teacher	Student	Student	Student

Source: Llewellyn, D. 2007. *Inquire within: Implementing inquiry-based science standards.* Thousand Oaks, CA: Corwin Press.

Enough Questions Already!

Depending on their experience in elementary school with science—and unfortunately, in many schools there's little time for science after all the math and literacy and test prep—a good number of your students may come to your class with apprehension and misconceptions that require you to evaluate and monitor their readiness to lead their own learning in doing good science. Research by Furtak et al. (2012); Duschl (2008); and Kirschner, Sweller, and Clark (2006) affirms that teachers need to adjust the degree of guidance they provide when conducting activity-based science according to the needs, abilities, and readiness of their students, as gauged by preassessment, assessment probes, and ongoing formative assessments (see Chapter 6 for more on good science assessment).

This is a call for differentiation in good science instruction (which we examine in Chapter 6). Beyond mere apprehension, for some students in the middle grades—depending on their cognitive development and comfort with hands-on, minds-on procedures—good science can be very difficult. Even when science is present in the elementary classroom, it is often characterized by direct instruction rather than hands-on, minds-on activities. We've found that middle schoolers whose previous science success emerged from parroting back factual information are prone to initial frustration with "doing" science.

Our goal is to bring students from their varied starting points to collaborative investigations and then to an understanding of scientific concepts. Depending on their science experience before your class, your students may be accustomed to memorizing answers, not generating and pursuing questions. Give it time.

Scientific Argumentation

Essential to thinking like a scientist and understanding the nature of science is engaging in scientific argumentation. In short, scientific argumentation (or simply "argumentation") is how students present and justify the claims they make from evidence, verbally and in writing, after analyzing their data. Here's a helpful way to explain argumentation to your students:

> [Scientific argumentation] is a bit like an argument in a court case—a logical description of what we think and why we think it. A scientific argument uses evidence to make a case for whether a scientific idea is accurate or inaccurate. For example, the idea that illness in new mothers can be caused by doctors' dirty hands generates the expectation that illness rates should go down when doctors are required to wash their hands before attending births. When this test was actually performed in the 1800s, the results matched the expectations, forming a strong scientific argument in support of the idea—and hand-washing! (Understanding Science Project 2012).

Doug Llewellyn and Hema Rajesh describe how argumentation fits into good science practice as "a progression where students (1) investigate questions and assumptions from a puzzling phenomenon or event, (2) use the data from a self-designed investigation to make a claim and justify and defend the claim with supporting evidence, and (3) provide a scientific explanation based on the findings" (2011, p. 22). Additionally, fellow students listening can and should ask challenging questions of presenters (See Figure 2.1).

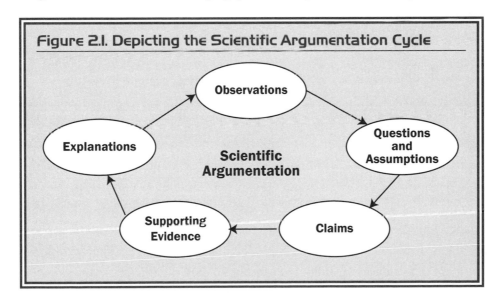

Figure 2.I. Depicting the Scientific Argumentation Cycle

Key words in depicting the important role of argumentation are *evidence* and *explanation*. A scientific argument must provide explanations emerging from and supported by evidence. Sometimes students (and teachers!) confuse the roles of data and evidence. "*Data* are the information and measurements from an investigation. *Evidence* is a particular subset of data an investigator uses to support or refute a claim" (Llewellyn and Rajesh 2011, p. 23; authors' emphasis).

In terms of thinking (and behaving) like a scientist to grasp the nature of science, even an activity driven by scientific inquiry or a well-designed engineering problem is incomplete if it simply arrives at more questions or a design solution.

The next step in good science is to go beyond scientific inquiry and engage students in scientific *reasoning*. "If scientific inquiry asks, 'What if . . .?,' then scientific reasoning asks, 'Why . . . ?'" (Llewellyn and Rajesh 2011, p. 24). Not only does argumentation take the activity to a more sophisticated level of thinking, it taps the middle schooler's natural desire to, well, *argue*. Middle schoolers *love* a persuasive challenge, and scientific argumentation challenges them to take a stand (based on the evidence, though, not just their opinions). Through argumentation, "students use scientific reasoning skills when explaining how the claim and the evidence are connected" (p. 24). See Figure 2.2 to see how scientific reasoning links claims to evidence and Figure 2.3 for teacher prompts to promote scientific argumentation.

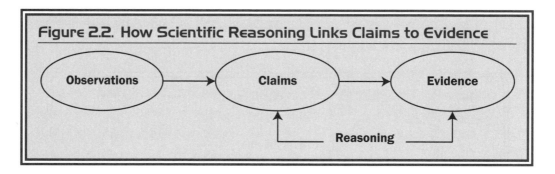

One of our goals in helping students think like scientists and attain scientific reasoning is conveying the concept that a scientific discovery is only meaningful if it is helpful in some way. "Hypotheses and theories live and die by whether or not they work—in other words, whether they are useful in explaining data, generating expectations, providing satisfying explanations, inspiring research questions, answering questions, and solving problems. Science filters through many ideas and builds on those that *work*!" (Understanding Science Project 2012). Scientific argumentation forms the connection between a question or problem and a meaningful, defensible, helpful solution, and builds in students an appreciation for the essential role evidence must play in making scientific claims.

Figure 2.3. Teacher Prompts to Promote Scientific Argumentation

- What assumptions can you make about the observation?
- What is the basis for your claim (or inference)?
- What evidence did you collect that supports your claim, idea, or hypothesis?
- Why do you think this is so?
- How does the evidence support or contradict your claim?
- What do you think the data imply?
- What conclusions or explanations can you draw from the evidence?
- Can you design a model to support your explanation?
- Were your original assumptions about the question or problem correct?
- How will you defend your findings?

Adapted from Llewellyn and Rajesh 2011, p. 25–26.

Arranging and Conducting Collaborative Table Groups

What do the *NGSS, Common Core State Standards, Mathematics* and *ELA*, 21st-century skills, 5E Model, and student engagement all have in common? If you get a vision of students putting their heads together and collaborating on an investigation, you are exactly right! Corporate America is telling us loud and clear that we need to shape students who think critically and creatively, communicate effectively, and collaborate with others. This rarely happens when students are seated looking at the backs of their peers' heads. Good science requires a different classroom arrangement.

The solution: collaborative table groups for your science students. This can entail either seating your students at actual tables of four, or having a quick way for students to turn their desks or bodies efficiently so they can form foursomes (our recommended group size) to begin exchanging ideas on topics that require critical and creative thinking. Any of you with elementary classroom experience already know that seating students in collaborative table groups develops their social and communication skills. In middle school, most of our students have no problem being social or communicating—our task is to focus them on *speaking cordially and purposefully* while *making scientific thinking visible*. Students doing good science must process what they hear and align it with their thinking in order to understand and apply it.

Take heart, those of you new to this challenge, because it can be done! Ideally, teachers have classroom furniture that includes lab tables, which double as desks, enabling students to sit and be able to converse together. These tables, or two pushed together, should be large enough for students to sit in groups of four. Teachers who only have individual desks can still have students learn to quickly move their individual desks to get in close proximity to three other students. We've seen this effectively done in a chemistry class where, on the teacher's signal, students quickly move their desks from rows to table groups.

Students should be strategically placed in groups. It may take a couple of weeks to get to know personalities and figure out who will work well together, so groups will be very fluid at first.

Optimally, groups will have diverse abilities and personalities that lend to the spirit of helpfulness and lively discussion. Seating students with varying abilities helps your students to be successful as they use each other's strengths to accomplish tasks. Keep in mind that the best reader should not be paired with the struggling reader in a group. In our experience, this makes things uncomfortable for both. Instead, pair an average reader with the struggler for the best results.

Before launching a table group activity, take time to have the students establish norms for group collaboration, and post them for reference and to help the students metabolize your shared expectations. One of the first expectations to establish when moving students into groups is that there should be no cheers or jeers as groups are made. Help lead the students toward this class norm before beginning your first table group activity. With their heightened self-consciousness and idealism, middle schoolers quickly decide together that embarrassing their peers is not okay, and if it's their decision rather than yours, you'll find it easier to enforce when a group strays from the rule.

Creating a culture of serious, collaborating scientists and a climate of respect and safety takes time and patience and will take practice on the part of the teacher and students. If we expect students to speak, listen, read, and write with one another, they need to feel safe and comfortable in the classroom. Let students know that it is a skill to be able to work

---

Figure 2.4. Sample Expectations for Table Groups

- Divide up roles and duties before beginning.
- Everyone does their fair share of the work.
- Listen completely before commenting.
- Begin each critique or constructive criticism with at least one positive comment.
- Ensure that everyone in the group participates.
- Expect respect and model respectful cooperation—walk your talk.
- Compromise when necessary to maintain forward progress.

cooperatively with other students in groups, and it is a privilege to be seated in a group that they earn by working at the skill together. Disrespectful behaviors that undermine a group member's confidence or willingness to participate erode the feeling of safety necessary for effective collaboration. See a partial list of group norms in Figure 2.4.

Granted, norms like the ones listed in Figure 2.4 don't always come naturally to middle schoolers, even when the teacher does a good job of implementing groups and engaging students in setting the expectations. Some of the group training needs to happen by working in groups, and working through issues together with the teacher when needed.

Teachers can expedite the effective function of their table groups by using specific, meaningful praise: "Catch them doing good" works very effectively with middle schoolers! When you see groups working well together, compliment them on the specific behaviors or qualities you appreciate. Let them know you appreciate their maturity (and score big points). Targeted praise early in the term will help establish effective table groups.

Once student groups have been established, number individual students 1–4 in the same position in each group. Facing a group of students, have the person to your right be #1 in each group, and count counter-clockwise one to four

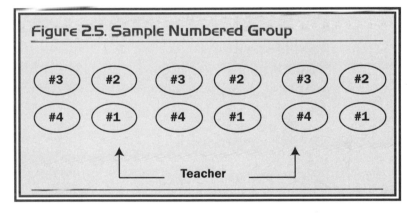

Figure 2.5. Sample Numbered Group

| #3 | #2 | #3 | #2 | #3 | #2 |
| #4 | #1 | #4 | #1 | #4 | #1 |

Teacher

(see Figure 2.5). This way, when something needs to be handed out or retrieved, you can call a set of numbered students to come get the items for their table groups. This keeps you from running around the room handing out materials, and saves a lot of time. (We sometimes use this as a management tool, too; if you have students who are especially fidgety and need to move more than others, give them the same number and get them up and moving, gathering papers, getting equipment, and so on without drawing attention to them.)

Once table groups are set, this is where the fun begins. Students begin talking to one another—about *science*.

Like any other skill you want your students to practice, you need to scaffold the experience and start with short, simple group tasks. At first, have students only talk with their shoulder partner about a very specific topic for a short amount of time (we recommend two minutes total, using a timer everyone can see to keep track), with

CHAPTER 2

specific objectives outlined for their exchange. This keeps things fairly controlled and helps students feel comfortable speaking out loud about the science topic or activity they're examining. Then, have students share across the whole group—for starters, we recommend structured short bursts of less than a minute for the whole group sharing.

As the students get better at interacting and staying focused on their topics, you can expect and require more in terms of productivity. Your role is to be diligent about circulating around the room, listening to conversations, redirecting where necessary, and complimenting groups that are focused on the task at hand.

Seating students in groups is also a natural fit for having students work cooperatively in lab activities and investigations. When you begin to introduce more complex tasks for the table groups, we've found it very helpful to give each student a role. Some of the most common roles we've used or observed, depending on the table group activity, include:

- Reader
- Recorder
- Reporter
- Monitor
- Time-Keeper
- Safety Officer
- Clean-up Leader

As needed, students can have more than one role—for example, the clean-up leader can also be the recorder.

Throughout the table group activities, build in time for students to take notes and write their observations or findings in their lab notebooks (see Chapter 3 for more on lab notebooks). To optimize the quality of these writings, we've found that when students know they'll be sharing what they write with others, it tends to up the ante. Build in time for reading aloud and exchanging writings within groups and/or with the whole class, as appropriate, and make this step a routine part of the table group process. The sharing tends to improve the clarity of student writing.

Once they're up and running, to structure collaborative table groups our rule of thumb is to shift group activities every eight minutes or so: writing in notebooks or reading short pieces on the topics, talking to shoulder partners, reading from their notebooks, talking in their group about the topics, writing responses to the sharing, and so on. The more students are writing, reading, thinking, and speaking about the science content they're examining together, the more conceptual understanding will take place.

Being Flexible and Fostering Cooperation

We've found collaborative table groups to be an optimal arrangement because like the majority of contemporary scientific endeavors (and the workplace of the future), good science activities are most often cooperative in nature. That said, we recognize Doug Llewellyn's sage observations about teacher flexibility in groupings:

> Not all students like to work collaboratively in groups. Some prefer to work autonomously and privately as individuals. Others are partial to working in pairs. Flexible groupings allow students to have a voice in selecting the group size in which they learn best. This includes a variety of individual and small-group settings, which matches their needs and learning styles. Flexible groups may include both teacher-assigned and self-selected work groups depending on the following:
>
> - the nature of the activity
> - the amount of classroom time available
> - the maturity of the class (or the individual)
> - which students can and cannot get along together
> - the extent of available supplies
>
> Although student choice is a main contributor to motivation [in good science settings], it is the teacher who is ultimately responsible for determining the size and configuration of the group. (2013, p. 57)

Granted, we know from experience that once in a while we all encounter a batch of kids who are simply *too* social to be productive in groups, even with excellent procedures—and we understand that teachers have to be creative with these occasional "special crops." Overall, though, good science at the middle school level incorporates various instructional tactics central to cooperative learning (as explained later in this chapter)—and it needs to, given the highly social nature of middle grades learners.

Teachers can foster cooperative skills—listening and respecting others, encouraging, explaining, summarizing, checking for understanding, and disagreeing in an agreeable way—by showing students that in their groups they must

- talk about the work,
- drill each other on the material,
- share answers,
- share materials, and
- encourage one another to learn.

There are many excellent examples of other cooperative tactics available to teachers in the professional literature, including several texts we list as resources in Chapter 8.

Collaboration: Write–Pair–Share

Another cooperative approach we find effective and readily adaptable to middle school classrooms is write-pair-share. Following an activity, demonstration, or discussion, students are asked to (1) think about what they just observed or talked about and (individually) write down two or three of the key ideas in their own words; (2) turn to their lab partners and listen to each other's ideas; and (3) move as a team to another group and repeat the sharing (or participate in a whole-group sharing). This strategy is both efficient and engaging.

Cooperative learning, however, is still perceived by some teachers as threatening because it requires them to relinquish some control depending on the lesson and activity. In conducting science and engineering practices, students are often up and moving, clustering in groups, comparing notes, negotiating, and generally making noise. For teachers to adapt to this classroom environment, they need a tolerance for ambiguity and a willingness to focus on student learning rather than on their own teaching as the primary force behind the lesson.

In our experience and in testimonials from middle-level teachers we interviewed for this book, we find that the number-one reason that middle school science teachers shy away from collaborative activities is the fear of losing control. So much of your success in middle school rests on establishing effective procedures within which that ambiguity, spontaneity, and energy can flourish—to a point. For example, to restore a conversation level, maintain control or signal clean up, we know teachers who establish a trigger signaling "Quiet, I need your attention please," or "clean up." The trigger could be flashing the lights, playing a certain song, or a catchy call-and-response; but any procedure is in lieu of shouting over the noise level. Kids will know what the signal means and react accordingly.

Cooperative science activities in middle school need not be chaotic or anarchic at all (and we think middle grades principals will appreciate that!). Teachers using good science methods effectively—with focus questions, established procedures, engaging activities, and clearly defined objectives—are not losing structure, they are changing the *type* of structure. Middle schoolers need a degree of predictable order, and inquiry activities can be readily arranged to meet this need, provided they are used in tandem with other instructional methods suited to the differing demands of the middle-level learner.

In addition to collaborative table groups, write-pair-share, and other cooperative strategies, scientific and engineering activities conducted by students individually can also be a component of the middle-level science teacher's repertoire, as can investigations based on a teacher demonstration of a lab activity, video simulations, and direction instruction. Compared with collaborative student exercises, these latter learning modes may not capitalize on the adolescent's need for social interaction, nor will they necessarily enhance the student's ability to define an independent role in an interdependent team—as a scientist does. But they may be very helpful for students whose learning style isn't immediately compatible with an interactive group approach and who need to be coached to work more comfortably in a group pursuing an open-ended problem or question. The key is identifying and adjusting one's teaching methods to differentiate student needs through ongoing formative assessments, observations, and interaction and finding the methods that work for different students and groups of learners. We'll say more about this in Chapter 6.

The point is, lessons can be adapted to involve more or less teacher control, affording tremendous flexibility. Through the use of procedures taught, tested, and reinforced during the year, middle-grade science teachers can lend structure to an activity by introducing a specific problem or topic, then allowing students the freedom to generate their own hypotheses and tests, control the project design, and work collaboratively toward project completion. In the pre-*NGSS* era, this blend of structure and freedom was referred to as *structured inquiry* and is an excellent match for the mix of order and spontaneity necessary to succeed with middle school learners. A structured inquiry complements effective cooperative learning—with its emphasis on individual accountability, interdependence, group interaction, and reciprocal feedback—and is compatible with collaborative table groups. It's a great way to introduce middle schoolers to doing good science.

On a practical note, we must stress that preparing for, setting up, and conducting activities incorporating science and engineering practices, with or without kits, takes time. We've found that while they are less intensive to set up, teacher demonstrations can be quite effective with middle schoolers as part of a mixed array of good science methods a teacher employs. Using the BSCS 5E Instructional Model from our activities, the teacher asks the focus questions and helps frame the problems, and students do the exploring, explaining, extending, and evaluating; this approach is a convenient and an adequate alternative in a time crunch (and it also cuts down on safety risks with younger middle schoolers).

We recommend that middle-level teachers present a blended array of demonstrations, short lectures, hands-on lab activities, and reflection exercises—in essence, a mix of teacher- and learner-centered instructional strategies. With this in mind, we move to Chapter 3, in which we present two teaching scenarios—from Mr. Hohum's class and Ms. Gottitrite's class—that illustrate the difference between traditional science teaching and what we consider *good science.*

References

Duschl, R. 2008. Science education in three-part harmony: Balancing conceptual, epistemic, and social learning goals. *Review of Research in Education* 32: 268–291.

Furtak, E., T. Seidel, H. Iverson, and D. Briggs. 2012. Experimental and quasi-experimental studies of inquiry–based science teaching: A meta-analysis. *Review of Research in Education* 82: 300–329.

Gerlach, J. 2012. Elementary design challenges: Students emulate NASA engineers. In *Integrating engineering and science in your classroom*, ed. E. Brunsell, 43–47. Arlington, VA: NSTA Press.

Kirschner, P. A., J. Sweller, and R. E. Clark. 2006. Why minimal guidance during instruction does not work: An analysis of the failure of constructivist, discovery, problem-based, experiential, and inquiry-based teaching. *Educational Psychologist* 41 (2): 75–86.

Llewellyn, D. 2002. *Inquire within: Implementing inquiry-based science standards.* Thousand Oaks, CA: Corwin Press.

Llewellyn, D. 2013. Choice: The dragon slayer of student complacency. *Science Scope* 36 (7): 90–95.

Llewellyn, D., and H. Rajesh. 2011. Fostering argumentation skills: Doing what real scientists really do. *Science Scope*, 34 (1): 22–28.

Pratt, H. 2012. *The NSTA reader's guide to* a framework for K–12 science education, *second edition: Practices, crosscutting concepts, and core ideas.* Arlington, VA: NSTA Press.

Understanding Science Project, The University of California Museum of Paleontology. 2012. The logic of scientific arguments, from *How science works. http://undsci.berkeley.edu/article/howscienceworks_07*

CHAPTER 3

What Good Science Looks Like in the Classroom

Imagine you have been assigned to record two middle school science classrooms, located side by side, covering the same topic (Newton's first law of motion, let's say) on the same day with comparable student groups. You will enter each classroom silently, without disrupting instruction, and record what you see to compare the two encounters later.

In Mr. Hohum's room, here is what you record:

- It is dark except for the glare of his projector.
- Students are sitting in straight rows, facing the teacher at the front of the room.
- Mr. Hohum is sitting next to his laptop, keeping one eye on the class while he flips through his lecture note slides.
- The classroom is quiet and orderly, and Mr. Hohum is the only one speaking.
- The teacher has defined Newton's first law and students are scribbling notes about objects at rest staying at rest unless acted on by a force.
- The lights go on and Mr. Hohum directs the class to read a section in their text and complete a worksheet on Newton's first law.

This scenario illustrates a teacher-centered lesson, with Mr. Hohum "telling" and the students passively receiving information. They are learning about science, as opposed to "doing" science. If the principal were to drop in on this lesson, she might admire the orderly conduct of Mr. Hohum's students, their attention alternately fixed on the slides, their note taking, and then textbooks and worksheets. Through exposure to this sort of teacher- and subject-centered routine on a daily basis, these middle schoolers have been sedated by inactivity and numbed by passive learning. Most of them probably aren't crazy about "science," either, thanks to Mr. Hohum's linear instructional methods. They have been programmed to "sit-get-spit-forget" whatever they're presented by the teacher, and, like Mr. Hohum himself, they unfortunately have counterparts in many middle school science classrooms across the country.

Stepping next door, your camera moves through the doorway of a very different learning environment in Ms. Gottitrite's class:

- The room is comparatively noisy, bustling with energy and simultaneous discussions at each separate station.

- Students are up and moving, working through procedures in small groups on two activities at different stations: a balloon car and a balls-and-ramps experiment designed to help students probe the relationship between force and motion and explore Newton's laws firsthand.
- Some students at the stations are recording data and others are discussing their observations, even disagreeing on key points; some are testing, some are analyzing, some are preparing their reports.
- The teacher is moving between groups of students, asking questions, challenging them to expand and apply their observations to possible explanations of the phenomena they saw. She is carrying a clipboard to note her own observations about levels of student understanding—areas where she may need to reteach because her students didn't "get it."

In Ms. Gottitrite's class, the students are "doing" science, and the teacher is prompting them to connect their observations to possible explanations, form hypotheses, experiment, gather data, and develop conclusions. This is learner-centered science, exemplifying the sort of classroom setting in which data collection, analysis, interpretation, argumentation, and other scientific practices (and 21st-century skills) flourish. Ms. Gottitrite's students are conducting experiments and building the experiential foundation on which concepts like Newton's first law can take hold when the class shifts to a more directed study of the history of and theoretical basis for this essential principle of physics.

This isn't to say there's no place for direct instruction methods, especially when a traditional lecture is needed to present challenging content. Furthermore, some students are conditioned by lecture and note taking, and they struggle in their transition to hands-on, minds-on instruction, even with teacher assistance. As a matter of well-rounded learning (following from the discussion in the previous chapter), students should be exposed to lecture and book work, and there are those learners who are best suited to direct instruction overall.

Also, because middle schoolers are so social, collaborative table groups just aren't productive at times even with the best procedures, and teachers must adjust their methods (and table group personnel) accordingly. Our observations of effective science classrooms suggest that good science instruction involves a combination of approaches, including active investigations in conjunction with direct instruction, independent research, and cooperative team projects.

Still, to accommodate different learning styles, instructional objectives, and developmental needs, the National Science Teachers Association recommends as a rule of thumb that 80% of time in middle school science classrooms be spent in active investigations (60% in the elementary grades and 40% in high school science; NSTA 2003).

The 5E Instructional Model

We recommend as a foundation the model of science instruction and lesson planning commonly referred to as the BSCS 5E Instructional Model, originating with Rodger Bybee and the collaborative team associated with BSCS (Bybee et al. 2006, 1990). Briefly summarized, the 5E components are as follows:

- **Engage:** The goal of this phase is to get students interested, helping them connect their prior learning (determined in the pre-assessment) to the new activity, focusing their attention on possible outcomes of the activity, and mentally engaging them in the practices and concepts to be studied. This phase launches the process.

- **Explore:** The goal of this phase is to get students involved in the activity or topic. Students actively interact with the new materials and process, conduct the investigation, develop the design solution, and build their own understanding.

- **Explain:** In this phase, students have opportunities to communicate their new understanding and demonstrate that they can perform the skills and understand the activity's concepts. In this phase, teachers introduce terms, provide explanations, and review concepts in small- and whole-group settings.

- **Elaborate:** In this phase, students apply their new skills, knowledge, and conceptual understanding to gain practice, broadening their understanding while pursuing areas of interest or challenge and refining their skills.

- **Evaluate:** Learners demonstrate their skills and understanding; teachers assess student performance of the activity's tasks and grasp of key concepts. Evaluation takes place at every phase of 5E as teachers watch and listen to students moving through the activity.

These are the five procedures we use in our sample activities that are time tested and proven across a range of school communities, student profiles, and subject matter.

Here's an illustration of how 5E might work in a middle school science class:

- The teacher determines a topic to investigate and provides a discrepant event or focus question or problem relevant to the students' past science experiences to *engage* student interest and curiosity. (Note that *students* might determine topics later in the year.)

- The students, with their teacher as a guide and co-investigator, begin to *explore* the problem or question.

- They make further observations and attempt to *explain* the phenomena they observe.

- The teacher then challenges students to *elaborate* on their understandings by linking observations to prior knowledge, supporting their conclusions with evidence, and applying their discoveries to the scientific concepts they're studying.

- Throughout the activity, and markedly in this phase, the teacher encourages students to *evaluate* their understandings and abilities, and the teacher *evaluates*, or assesses, the areas of strength and weakness exposed by student performance in the activity.

As part of the evaluation, the class reflects on where the investigation can go next: What new questions do students have that can be pursued? How might the problem be framed differently, and might that lead to a different outcome or finding? If the investigation failed, what variables need to be addressed to make it work and try again? In these sorts of reflective and critical outcomes, the 5E model helps students "see their thinking" and practice thinking like scientists.

Note that every component of the 5E model need not be covered in a single period. It could extend over several days. A sixth-grade teacher in a self-contained classroom could do *Engage* in the last 15 minutes of the school day, cover *Explore* the next day, and so forth. Deploying the 5E model in this way enables time-crunched teachers to "sneak" good science in even if they don't have a full hour to devote to an activity or investigation.

Other important aspects of the 5E model are evident in Ms. Gottitrite's teaching vignette and in the activities we provide for you. The 5E Model and student-centered strategies serve middle school science teachers and students by providing a rich context for relevancy. According to master middle grades science teachers we interviewed for this book, one of the most important reasons to incorporate the 5E Model into a broad repertoire of teaching tools is that the method helps students make connections between conceptual science and experiential (real-life) science understandings. As Bruce Jones, Mesa's secondary science specialist, explains, "I feel like when my students have the chance to really experience science through the 5Es it becomes personal to them. It is no longer about what Mr. Jones says, but about what they have come to understand for themselves." By "doing" science, students develop a set of experiences that add meaning and essential relevance to the scientific concepts and principles delivered in lectures and textbooks.

Learning Science as a Process

In the process of doing good science, students acquire valuable skills through their interaction, collaboration, and problem solving with other students—skills that cannot be learned sitting at desks in rows and listening to a teacher. Experiencing science as a process, rather than a database of facts, is analogous to learning to play a musical instrument; we don't expect students to pick up piano by reading about it or watching the teacher play. These activities might be part of the overall strategy, but the students have to "do" piano in order to play it, just as middle schoolers must "do" science to master its concepts.

This is especially true in the digital age. We know from Yong Zhao (2012) and others who've studied the misalignment between the needs of the 21st-century workforce and the current, 19th-century approach to educational "reform"—emphasizing acquisition of content, standardization of instruction, and homogenization of curriculum—that students today have unlimited access to information. Indeed, they can "know" (as in conduct an internet search for) everything their teachers know. If the role of teachers used to be helping children acquire knowledge to prepare them for the future, our job is to help our students make meaning out of the endless information that bombards them, developing their skill to filter fact and opinion, relevant and irrelevant information, evidence and rhetoric.

To look more closely at what we mean by good science, let's examine the NRC's statement on the nature of science in the *NGSS*:

> Suppose students observe the moon's movements in the sky, changes in seasons, phase changes in water, or the life cycle of organisms. One can have them observe patterns and have them propose explanations of cause-effect. Then, have the students develop a model of the system based on their proposed explanation. Next, they design an investigation to test the model. In designing the investigation they have to gather data and analyze data. Next they participate in the practice of constructing an explanation using an evidence-based argument. (NGSS Lead States 2013, p. 7)

Note the key *action* verbs in this passage—science is *observing, proposing, explaining, modeling, designing, investigating, testing, gathering and analyzing data,* and *constructing arguments*—all active, all dynamic, all procedural. Good science is something students *do*, not something that's done to them. In hands-on, minds-on science, students become "co-investigators" rather than vessels for information; they share with their teachers the responsibility for engaging in the process, the outcome, and the reflection during an investigation, experiment, or field study. This is part of a larger structural shift in our approach to science instruction in the middle grades (see Table 2.1, "Big Shifts in Science Education," on page 13).

Just as we urge that teachers and administrators collaborate to make these sorts of changes possible, we also believe in consulting the ultimate authority when it comes to determining what works and what doesn't: the students.

Student Response to Good Science

While educators traditionally haven't involved students in instructional decisions, we have observed over the past four decades that the reaction from students engaged in good science across the United States has been resoundingly enthusiastic. Quite simply, when students (and particularly middle school students) have an opportunity to engage actively, collaboratively, and frequently in the learning process, science becomes their favorite subject.

Let us illustrate this student enthusiasm for good science with a case example from our own district (the Mesa, Arizona, public schools). Hands-on, minds-on science engages our middle schoolers such that several years ago, when science was inadvertently listed as "optional" on a seventh-grade preregistration form, less than 5% of students at a feeder school then using traditional text-based approaches signed up for science—compared with more than 95% of the sixth graders at a different feeder school with an inquiry-based program!

Those of you who are already "living" good science with middle schoolers have no doubt heard student comments such as these provided by teachers we consulted for this book:

- (start of lab): "Oooh, this is gross!" (same student, end of lab): "I think I want to be a vet."
- "My brain hurts."
- "I hate you, Ms. Gratkins. You make me think!"

About the *NGSS*

The *National Science Education Standards* were viewed as progressive and forward-looking a decade ago and are still fundamental to our understanding of good science. Yet no one could have anticipated the impacts of globalization, the digital revolution, or online learning as they have reshaped the aims of schooling worldwide.

Recognizing the urgent need to re-evaluate the aims of science education, in 2009 the National Research Council (NRC) and the National Academy of Science's Board on

Science Education, Division of Behavioral and Social Sciences and Education, convened a task force to re-examine what and how science should be taught in America's schools. A diverse group of educators, scientists, and researchers composed this task force, charged with recommending what content and pedagogy would form the foundation for new K–12 science standards.

Armed with major studies and new research on how children learn science, and considering the role science plays in the *Common Core State Standards*, STEM education, and the 21st-century skills for students, the task force generated *A Framework for K–12 Science Education: Practices, Crosscutting Concepts, and Core Ideas*, published in 2011 by the National Academies Press. The *Framework* became the basis for the *Next Generation Science Standards* (*NGSS*), which emerged in 2013 after extensive public review and revision. (Co-author of this book Vicki Massey served on a state advisory committee that helped shape the creation of the *NGSS*.) States not officially adopting the K–12 standards are still using the *Framework* and *NGSS* to guide best practices in their respective states.

So now good science in your classroom has a new template in the *NGSS*. At first glance, we know that some teachers find the *NGSS* complicated and unintuitive. Many excellent science teachers are comfortable with the content they've mastered and the methods that work with middle schoolers and believe that initiatives in our crazy profession come and go all the time. Are the *NGSS* just the latest mandate passing through on the way out?

Let's stop right here and think about evolution as we teach it to our students, as natural selection. Imagine an animal that does not adapt to the changes in its environment. In the face of irreversible forces acting against it, the creature struggles to resist, perhaps modifying its behaviors or habits, gradually experiencing vulnerability and misery, and ultimately dying. Hyperbole this may be, but your authors are determined to help steer you away from unnecessary misery!

So, let us help you adapt to the *NGSS*. We'll take it slowly and deliberately, like the evolving animals we are.

Good Science and the *NGSS*

Understanding the *NGSS* is the first step toward appreciating their usefulness. At first pass, the *NGSS* appear markedly different from past standards, but we think that's a good thing—once you become accustomed to the formatting, the *NGSS* are more at-a-glance-comprehensive, actually listing *Common Core* language arts and math standards connections at the bottom of each concept, for example. After perusing the document a few times purposefully, you'll begin to see the value behind the formatting. We'll help.

Two major goals in the *Framework* are to (a) educate all students in science and engineering, and (b) provide the foundational knowledge for those who will become the scientists, engineers, technologists, and technicians of the future (Pratt 2012, p. 9). Numerous studies and reports, including *Prepare and Inspire* (PCAST 2010), point to the central role that STEM (science, technology, engineering and mathematics) education will play in America's economic growth and prosperity—including STEM's reach into "everyday" jobs as well as the rapidly expanding high-tech sector.

While the future is uncertain, it is a safe bet that most jobs will on some level become STEM jobs. We realize that K–12 education is more than workforce development, yet we believe it imperative for teachers at every grade level and especially in the secondary divisions to recognize our responsibility to shape instruction in ways that are relevant to the lives (and future livelihoods) of the students we serve. (See Appendix A for more on teaching about STEM careers.)

The *NGSS* aim to promote depth over breadth, provide a more articulated continuum of learning across the grade levels, and help students understand the "crosscutting concepts" that permeate the science disciplines. The new standards were designed to encourage application through inquiry and engineering design so that students can grasp the nature of science and its core themes. One major change is that every standard includes one or more engineering performance expectations.

The biggest difference between former science standards and the *NGSS* is the emphasis on *doing* science in the new structure—and the expectation that students be able to demonstrate their understanding through *performance*, rather than recall of facts. We know that children learn science best by doing it; the future demands employees who can *use* information, not merely memorize and repeat it. This, we know, is an outcome of good science, benefiting students in classrooms now and preparing them for their roles in the workforce of the future, so we see the *NGSS* as a powerful initiative in support of good science reforms.

In the *NGSS*, students demonstrate understanding by developing and constructing models, planning and carrying out investigations, constructing explanations, analyzing and interpreting data sets, providing evidence, using argumentation to substantiate findings, using maps and other visualizations, and more. For experienced middle school science teachers, the *NGSS* present somewhat different language than that with which we've become comfortable: explaining, describing, distinguishing between, recognizing, comparing, and so on.

Veteran science teachers, rest assured. The *NGSS* call for evolution, but not revolution—with a few tweaks to our beloved and familiar curricula, semantics, and pedagogy, and a willingness to try on some new approaches and activities like the ones we include in this book, you can definitely evolve with the *NGSS*.

Understanding and Applying the *NGSS* to Your Classroom

First of all, the *NGSS* are meant to be accessed online so that the formatting—color coding and links to click—can enhance your use and understanding of the electronic document. If you are working from a hard copy of the *NGSS*, you'll likely have a harder time than if you work through them online.

The *NGSS* are composed of three main elements: science and engineering practices, disciplinary core ideas, and crosscutting concepts. As we explained in Chapter 1, science and engineering practices are the skills students need to develop to be able to "do" science and engineering. The practices are the mechanisms for varying degrees of student-directed discovery, autonomy, and heightened engagement in science and engineering activities. Using these skills, students acquire discrete content (called *disciplinary core ideas* in the *NGSS*), and make connections across science disciplines via the crosscutting concepts.

At the top of the standard, delineated with small letters, are the *assessable components* or "performance expectations"—what students who demonstrate understanding are able to *do*.

Following each lettered assessable component/performance expectation, in smaller red font, is an "assessment boundary" that lets teachers know where to stop providing information relating to the concept. The assessment boundary identifies what information goes beyond the content and degree of complexity teachers need to provide students at each respective grade level. For example, students who can explain tides, seasons, eclipses and lunar phases don't have to know Kepler's laws of orbital motion—that's the assessment boundary for this *assessable component*. The assessment boundary gives us permission not to assess items with which students needn't have familiarity at their grade level.

Below the *assessable components* are the foundation boxes, in pastel colors: *science and engineering practices, disciplinary core ideas,* and *crosscutting concepts.*

Dimension 1 of the *NGSS* foundation boxes is *science and engineering practices.* This actually replaces what we might have referred to as *process skills* in the past and even encompasses the concept of "inquiry" that shaped the *NSES*, and to which many of us veterans became so partial.

According to the *Framework*, science and engineering involve eight critical "practices" outlined in this dimension, which support the assessable components:

- Asking questions (for science) and defining problems (for engineering)
- Developing and using models
- Planning and carrying out investigations
- Analyzing and interpreting data
- Using mathematics and computational thinking
- Constructing explanations (for science) and designing solutions (for engineering)
- Engaging in argument from evidence
- Obtaining, evaluating, and communicating information

Dimension 2 of the foundation boxes lists the *disciplinary core ideas: physical, life,* and *Earth and space sciences*. All of these *core idea* areas encompass fewer topics than in past standards, which makes them easier to manage.

As is the case with math and reading curricula in the past decade, younger students are expected to manage science concepts earlier than past generations did. Consequently, you'll note that some topics in the *NGSS core ideas* are expected to be covered at lower grade levels than in prior standards. Teachers and curriculum specialists will need to modify their curricula and assessments to reflect these changes—and this is part of that "evolution" we described above.

Dimension 3 of the *NGSS* foundation boxes presents the *crosscutting concepts*. This section should be familiar to teachers accustomed to the *NSES* (the unifying concepts and processes) and the *Benchmarks* (the common themes). The *crosscutting concepts* are just what the name implies—fundamental ways of knowing and organizing scientific processes that connect and intertwine through scientific ideas, topics, and practices:

- Patterns
- Cause and effect: Mechanism and explanation
- Scale, proportion, and quantity
- Systems and system models
- Energy and matter: Flows, cycles, and conservation
- Structure and function
- Stability and change

The three broad *NGSS* dimensions—science and engineering practices, disciplinary core ideas, and crosscutting concepts—correspond to one or more of the letters representing the *assessable components* (or performance objectives that students need to be able to accomplish). When teachers cover a performance objective, they can see exactly which of the three dimensions they're addressing. It's all on one page and very convenient, once you learn how to navigate the new format.

Science teachers as well as those responsible for math, social studies, and other subjects are expected to design lessons integrating the *Common Core State Standards, English Language Arts* and *Math* as well. Good science naturally integrates literacy and mathematics; conveniently, the ties to the *Common Core State Standards* are listed for us at the bottom of each concept page in the *NGSS*.

Once you become familiar with the *NGSS* format and structure, it's evident that the standards are rich with good science essentials. Just like the *NSES* that preceded them, the *NGSS* call on teachers to help children think like scientists and to *do* science, as it is conducted by professional scientists and engineers today. And we know that what the *NGSS* envision stands in contrast to much of what is presented to middle school students in science classrooms across the country. Some of this is because of the pressures wrought by a perfect and ongoing storm of math and literacy test preparation, accountability, reduced funding, and the general de-emphasis of the nontested subjects. These are tough times for good science.

And yet we also know (and have widely observed) districts, schools, and classrooms where the potential for good science is limited primarily by the teachers' familiarity with, interest in, access to, and willingness to try the methods put forth by this book and the *NGSS*. Using the *NGSS* effectively to guide good science instruction is one leg of a three-legged stool: content, standards, and methodology. The stool is our metaphor for the elements that must be intact for good science improvement in the middle school grades (beyond critical variables suchg as de-emphasizing rote standardized testing and addressing the crippling effects of poverty).

We believe that the hurdles facing middle-level science reform involve improving methodology as much as enriching teacher content knowledge, restoring support for good science as a core subject, and reforming preK-to-PhD science curricula.

Fundamentally, in the context of schooling, good science is inseparable from good teaching.

It's Good Teaching!

In Yong Zhao's *World Class Learners* (2012), a powerful critique of high-stakes testing and America's educational "reform" movement, he describes an ambitious educational model that would properly prepare our children for the jobs we haven't invented and the problems we can't yet imagine. Key among the principles of his new education paradigm are shifting responsibility for learning to the learner; offering students choice and agency; making them partners with teachers in their educational experiences; and involving them in collaborative design, problem solving, decision making, and investigation.

While Zhao's vision may be distant in the future for the vast majority of America's public schools, consider how good science is providing a vehicle to achieve many of his aims *today*.

Sometimes, we realize, a school or district curriculum or a high-stakes test battery truly dictates the nature of science instruction in a school; but in many cases, teachers with permission to adjust will not (or cannot) do so. And too often in middle schools, students are actually falling asleep in their textbooks.

Implementation Challenges and Subtle Shifts

While we are convinced—and our experience is reinforced by teachers across the country—that good science is ultimately easier, much more valuable and rewarding, more fun, and more cost-effective than traditional text-centered approaches, we would be remiss if we didn't acknowledge some of the challenges teachers can experience in implementing science and engineering practices.

For example, there never seems to be enough time to explore topics deeply and still cover the required curriculum. (Among its many revisions, the *NGSS* cover fewer concepts to enable more in-depth exploration of topics.) Teaching active science requires making time for preparation, setup, and cleanup (until students are trained to expect and follow procedures); transitioning from Hohum's traditional lessons to incorporate Gottitrite's good science strategies; implementing cooperative tactics like collaborative table groups; and assimilating varied formative assessment techniques. All of these objectives are more challenging in districts where science class sizes have swelled due to cost-cutting measures linked to state budget cuts.

In addition, the purchase, care, refurbishment, and storage of materials required for hands-on, minds-on science can pose logistical problems. Furthermore, good science units can be expensive (although they don't have to be—after start-up costs, consumables are cheaper than workbooks and reams of copy paper for worksheets). Some of us who are accustomed to teacher-centered methods struggle with the shift to being a "co-investigator"—in particular, the management challenges in training students to adjust their behavior when class activities switch from direct instruction to hands-on activities, and back again. Furthermore, helping students, colleagues, administrators, and parents embrace science and engineering practices and inquiry requires no small amount of strategy, foresight, and planning.

We mention these challenges not to dissuade or discourage readers but to help you prepare in advance to move gradually, adapt incrementally, and focus on short-term, achievable goals—given your comfort level and experience. The transition to good science in middle school involves subtle shifts rather than sweeping changes. Many teachers have reported to us that once students are in the good science groove—anticipating procedures, cooperating in setup and cleanup, and feeling comfortable about working independently of the teacher's constant guidance—activities are actually less work and far more rewarding than traditional "sit-get-spit-forget" science.

In terms of the perennial concern with funding active science, years of budget data from our own district resource center in Mesa are proof that good science is cheaper in the long haul than textbook-based programs (if you consider the hardware costs of e-textbooks in districts that now use them).

Considering the challenges and assets of good science, we understand that change is daunting to many of us who are struggling to meet daily objectives and teach well with multiple demands on our energy and time. That's why we advocate subtle shifts, which translate, practically, to developing and implementing a couple of new units each year. Trying to do too much too quickly is counterproductive for you and for your students.

References

Bybee, R., et al. 1990. *Science and technology education for the middle years: Frameworks for curriculum and instruction.* Andover, MA: National Center for Improving Science Education.

Bybee, R., et al. 2006. The BSCS 5E instructional model: Origins and effectiveness: A report prepared for the office of science education, National Institutes of Health. Colorado Springs, CO: BSCS.

Marzano, R., D. Pickering, and J. Pollock. 2001. *Classroom instruction that works: Research-based strategies for increasing student achievement.* Alexandria, VA: Association for Supervision and Curriculum Development (ASCD).

McComas, W. 2008. Seeking historical examples to illustrate key aspects of the nature of science. *Science and Education* 17 (2): 249–263.

National Science Teachers Association (NSTA). 2003. Position statement: Science education for middle level students. Arlington, VA: NSTA.

NGSS Lead States. 2013. *Next Generation Science Standards: For states, by states.* Washington, DC: National Academies Press. *www.nextgenscience.org/ next-generation-science-standards*.

Pratt, H. 2012. *The NSTA Reader's Guide to* A Framework for K–12 Science Education: *Practices, crosscutting concepts, and core ideas.* Arlington, VA: NSTA Press.

President's Council of Advisors on Science and Technology (PCAST). 2010. *Prepare and inspire: K–12 science, technology, engineering, and math (STEM) education for America's future. Report to the President. www.whitehouse.gov/sites/default/files/microsites/ostp/pcast-stemed-report.pdf*

Zhao, Y. 2012. *World class learners: Educating creative and entrepreneurial students.* Thousand Oaks, CA: Corwin Press.

Integration Is Key

Science, Technology, Engineering, Math, and Literacy

STEM: The Crossroads of Curriculum

How do we integrate reading, writing, and math with science? What is the appropriate role of technology in science and engineering practices? What does a STEM lesson look like? Good science instruction is by nature cross-disciplinary, weaving literacy and numeracy with problem solving, discovery, and other higher-order thinking skills. In so many ways, good science is the crossroads of curriculum.

This fundamentally integrated nature of science is manifest in the *Framework*, which puts forth the eight essential "practices" that comprise science and engineering design:

1. Asking questions (for science) and defining problems (for engineering)
2. Developing and using models
3. Planning and carrying out investigations
4. Analyzing and interpreting data
5. Using mathematics and computational thinking
6. Constructing explanations (for science) and designing solutions (for engineering)
7. Engaging in argument from evidence
8. Obtaining, evaluating, and communicating information. (NRC 2012)

The narrow focus on memorization and recall and increasing fragmentation of academic subjects brought about by high-stakes testing in the past decade—and notably the separation of science from math and literacy—is at odds with the call for more and better STEM education.

Indeed, the demand for increased and improved integrated STEM education is everywhere today, heralded loudest from the private sector in its ravenous hunger for STEM talent. The need is frequently cast in dire terms; that the current and future economy depends on a workforce with STEM fluency seems widely accepted as we write this second edition of *Doing Good Science* (see Appendix A for more on this topic).

The National Academies' 2007 report *Rising Above the Gathering Storm* made it clear that unless America advances its STEM education significantly, we can expect to forfeit our historical prominence in these fields to those countries that do so: "China (20%

of the world's population) and India (15% of the world's population) are challenging U.S. dominance in science and technology. India has nearly as many young professional engineers as the United States and China has more than twice as many" (Wojnowski, Charles, and Warnock 2012, p. 63).

We believe science teachers, of all people, acknowledge the urgent need for and value of STEM education. But what do we mean by STEM? Clearly, innovation in today's knowledge economy depends on each of the STEM disciplines, with communication and design as the connective tissues, yet a universally accepted definition of STEM is elusive.

> As educators, we seem to consider STEM singularly from an educational perspective in which success in science and mathematics is increasingly important and technology and engineering are "integrated" when appropriate. When you start to divide STEM by subject (the silo approach), it gets even murkier. Can science and mathematics alone be STEM? Does using an electronic whiteboard during a lesson make it a STEM lesson? When my kindergarteners are playing with building blocks, is that a STEM center? If you ask 10 different science, mathematics, technology, and engineering teachers to define STEM, each will give you a very different and unique answer. (Gerlach 2012)

We found this to be very true. Further questions arose when we talked with teachers about what they believe STEM embodies, beyond the call for workforce development in a rapidly changing world:

- Does an activity count as STEM if only one or two of the areas are engaged?
- How is STEM different from the integration we've been doing for years?
- What qualifies (or disqualifies) a science activity or investigation as "STEM"?
- Do science and mathematics take precedence as STEM content?
- Since STEM is often cited as critical to jobs and the economic future of America, how can teachers make the connections between STEM lessons and real-world applications in STEM careers?
- What *are* the STEM careers?
- How can an effective STEM curriculum, unit, activity, or investigation be assessed?

For the purpose of this book, STEM is defined as it is characterized in the *Framework*:

> In the K–12 context, "science" is generally taken to mean the traditional natural sciences: physics, chemistry, biology, and (more recently) Earth, space, and environmental sciences. … We use the term "engineering" in a

very broad sense to mean any engagement in a systematic practice of design to achieve solutions to particular human problems. Likewise, we broadly use the term "technology" to include all types of human-made systems and processes—not in the limited sense often used in schools that equates technology with modern computational and communications devices. Technologies result when engineers apply their understanding of the natural world and of human behavior to design ways to satisfy human needs and wants. (NRC 2012, pp. 11–12)

We'd characterize a lesson as a STEM science activity if it includes at least two of the STEM disciplines, and this was our approach with the sample activities in this book. We also like this STEM definition from Vasquez, Sneider, and Comer (2013):

STEM education is an interdisciplinary approach to learning which removes the traditional barriers separating the four disciplines of science, technology, engineering, and mathematics, and integrates them into real world, rigorous and relevant learning experiences for students (p.4).

Such STEM definitions are helpful, and for those of you already familiar with the principles of hands-on, minds-on science, they should strike you as *very* familiar! Now that STEM has entered the mainstream jargon of educators, you may find that you've been a STEM teacher for a long time.

To us, the emphasis on STEM seems to be a logical linear extension of the efforts driving good science since Highline, Anchorage, Fairfax County, and Mesa school districts developed the first science kits more than four decades ago. Indeed, the NRC characterizes effective STEM education in a manner that echoes the description of good science across these 40+ years: "Effective instruction capitalizes on students' early interest and experiences, identifies and builds on what they know, and provides them with experiences to engage them in the practices of sciences and sustain their interest" (NRC 2012, p. 18).

The *NGSS* reflect the collective call for STEM education as the evolution of inquiry-based science emerging from the 1970s and shaped by the *NSES* that preceded this iteration of the standards.

What is different in Next Generation Science Standards (NGSS) is a commitment to fully integrate engineering design, technology, and mathematics into the structure of science education by raising engineering design to the same level as scientific inquiry when teaching science disciplines at all levels from kindergarten to grade 12. This new integrated approach to science education is sometimes referred to by the acronym STEM." (NGSS Lead States, 2013; see Appendix 1)

In some regards, integration of the STEM disciplines is a must today, since "the interconnectedness of science and technology has resulted in a greater emphasis on the integration of engineering into the teaching of science" (Moyer and Everett 2012, p. 4).

Vasquez, Sneider, and Comer (2013) translate the *NGSS* call for integration in their clear and concise STEM guiding principles, summarized here by Rodger Bybee in his foreword to the book:

1. **Focus on Integration.** Combine two or more of the STEM disciplines so students can see the relationship among concepts.
2. **Establish Relevance.** Help the students develop meaning through the application of STEM knowledge.
3. **Emphasize 21st-Century Skills.** Help students develop the knowledge and skills they need for the contemporary workforce.
4. **Challenge Your Students.** Provide projects, tasks, and activities that hold students' interest and challenge their understanding and abilities.
5. **Mix It Up.** Provide a variety of STEM lessons and activities for the students. (pp. x–xi)

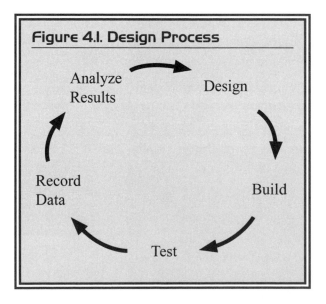

Figure 4.1. Design Process

Analyze Results → Design → Build → Test → Record Data → Analyze Results

Beyond the *NGSS* and K–12 classrooms, STEM is widely associated with workforce development, as we explore in detail in Appendix A (p. 215), including information to help you teach about STEM careers. Truly, middle school science is an optimal portal (and perhaps the last, best chance) to get students interested in STEM careers.

Engineering Design: Inquiry's Partner in Doing Good Science

Now then, let's address the elephant (with the slide rule—oops, graphing calculator) in the room. The emergence and prevalence of engineering in the *NGSS* may be intimidating to those of us whose academic background lacks grounding in physics and advanced mathematics to tackle college-level engineering problems ourselves. However, it's helpful to view "engineering" (rather, engineering *design*) as a *process* rather than a discipline when you envision the work of middle schoolers tackling a STEM activity (see Figure 4.1, previous page; adapted from Brunsell 2012).

The *NGSS* reflect a *progression* of capabilities and a developmental approach, summarized this way in the *Framework:*

> Children are natural engineers. They spontaneously build sand castles, dollhouses, and hamster enclosures, and they use a variety of tools and materials for their own playful purposes ... A common elementary school activity is to challenge children to use tools and materials provided in class to solve a specific challenge ... Middle school students should have opportunities to plan and carry out full engineering design projects in which they define problems in terms of criteria and constraints, research the problem to deepen their relevant knowledge, generate and test possible solutions, and refine their solutions through redesign." (NRC 2012, pp. 70–71)

The *Framework* calls out two core engineering teaching standards that guide our work with students in the "engineering design" dimension of STEM:

- ETS 1: How do engineers solve problems?
- ETS 2: How are engineering, science, technology, and society interconnected?

Both of these core ideas invite teachers to do good science—implementing active investigations that integrate multiple disciplines and enable students to see relevant connections to their world. In this sense, the increased emphasis on engineering design is really just an extension of the fundamentals of good science education that many of you are already doing.

If you are apprehensive about teaching engineering on top of everything else, you're not alone. But "the inclusion of engineering concepts and practices in the *Framework* is not intended to add more to the plate of teachers with an already overburdened science curriculum. Instead, these core ideas and practices are meant to help teachers introduce the interdependence of science, technology, and engineering and to harness the power of design activities to support authentic learning of science concepts" (Brunsell 2012,

> **Figure 4.2. Topics to Consider for Design Challenges**
>
> - Creating better levies: Coastal erosion
> - Building a better lunar lander
> - Creating the best reentry vehicle from space
> - Designing the best shipping container
> - Creating the best cooler to keep drinks cold in the summer
> - Designing an animal to live in a specific environment
> - Creating the perfect planter

p. x). Best of all, engineering design activities are hands-on, engaging, and fun—and will not require you to have a degree in physics to implement in your classroom. (See Figure 4.2, previous page; adapted from Brunsell 2012)

To understand the relationship between engineering design and scientific inquiry, recognize that while they are interrelated (as above), they are not the same. "The purpose of scientific inquiry is to use evidence to explain the natural and designed world. The purpose of engineering is to solve specific problems related to needs and wants" (Brunsell 2012, p. 3). Like inquiry, where questions lead to more questions, engineering activities are not necessarily linear (and can be "messy") because "failure is a constant companion in the design process" (Brunsell 2012, p. 3). Indeed, as Gerlach observes, "Students need more than one chance to be successful at a task. So many times they are left thinking, 'next time I would have …' The design process allows students to have that next time" (Gerlach 2012a, p. 43).

There are a numerous examples of engineering design process models circulating in various writings today; all offer a template that teachers can modify to meet the learning profiles of their students and the dynamics of certain activities. Figure 4.3 presents one model that we find easy to understand and use; we'd suggest adding audience participation to the argumentation process. Questions and challenges from classmates fosters good listening skills and increases everyone's attention to making reasonable assertions.

The Engineering Design Fair

Science fairs have become a staple of middle school programs nationwide. Traditional science fairs take a variety of forms and are more or less successful (and authentic) depending on the extent to which projects are completed by students rather than their parents. There's also a proliferation of science fair project ideas online that diminish the creative and critical thinking aspects of the process for children. The notion of individual students working on independent projects runs counter to the emphasis good science puts on collaboration.

Some schools are transitioning to a "design fair" model, showcasing the engineering design process outlined above, and reflecting the changing emphases emerging from the *NGSS* and the 21st-century skills. Gary DeMoss, a master science and technology teacher in California, advocates these basic components of a successful design fair:

1. Students form design teams (sometimes the same lineups as the in-class collaborative table groups) and develops a set of problems, differentiated according to each team's learning profile.

Figure 4.3. The Engineering Design Process

Stage I: Introduce the Problem

- Students and teacher prepare (or teacher introduces) a problem within a "design brief," including the goal of the project, expectations, and limitations (or specifications).
- Students conduct research about the problem and what design solutions may have been used in the past.

Stage II: Develop Possible Design Solutions

- Students brainstorm as many possible design solutions as they can (rapidly generating ideas without passing judgment).
- Complex problems may be divided into smaller pieces, and students brainstorm solutions to each piece.
- Brainstorming should be collaborative in table groups or whole-class rather than independent.
- Teacher keeps the focus on generating rather than evaluating design solutions.

Stage III: Analyze Design Solutions

- After generating a large number of possible design solutions, students begin analysis of how each solution can address the goals and specifications presented in the design brief.
- Students develop (or teacher provides) a systematic approach to their analysis, such as a simple matrix, to summarize pros and cons of each solution.
- Students select one or more solutions to prototype and test.
- Design solutions should include a description of how the solution works, how it matches the problem definition, and clear drawings or illustrations.

Stage IV: Test, Evaluate, and Revise Design Solutions

- Students collect data on the performance of their solutions and identify ways to improve them through testing.
- "Fair tests" change a single aspect of the design to evaluate how the change impacts performance.
- Teacher reinforces the concept that engineering design is iterative—whenever possible, testing should lead students to refine their design repeatedly (multiple retests are time-consuming but more authentically replicate the engineering design process that strives for continual improvement).

Stage V (and Throughout): Communicate

- Throughout the engineering design process, students should have opportunities to communicate their results and share their thinking with peers and, if possible, the entire class.
- Students can share the results of their tests, describe decisions they made in optimizing their solutions, and present the evidence supporting their decisions (argumentation).
- As a culminating activity, students develop a presentation of their design solution, evidence of success, and the path they took to solve the problem.

Source: Adapted from Brunsell 2012, pp. 3–4.

2. Design teams work together in class to do research and develop possible solutions. This is a very important step for research teams. It forces them to work together in class and spend out-of-class time collecting necessary materials to conduct their investigations. It also makes it easier for the teacher to monitor their ability to collaborate and assess their progress. (And it cuts down on parental micromanagement, too.)

3. Design teams analyze, test, and revise solutions in class.

4. Design teams present their designs to peers (in some cases, large student groups, younger grades, etc.) and prepare for the after-school design fair event.

5. Presentations emphasize process rather than product or solution.

The design fair could be adapted to or paired with topics emerging from inquiry-based activities, with the emphasis on collaborative teams, critical thinking and problem solving, and creativity—pulling students away from the tired science fair staples (e.g., teeth in cola, watering seeds with vinegar, magnets, and the like).

Putting the *T* in STEM

Unquestionably, the digital explosion put technology at the forefront of K–12 schooling, and science teachers benefited from the proliferation of new tools. Through laptops and tablets, interactive projection, probeware, digital media, and other emergent innovations that may make all of this obsolete by the time you read this book, technology is driving budgeting priorities, if not necessarily instructional improvement, in schools across the nation.

But what do we mean when we refer to "technology"? To what extent does technology extend STEM education? Is good science dependent on technology, or independent of it?

We refer to the definition of STEM from the *Framework*: "Technology" is what results when we "design ways to satisfy human wants and needs" (NRC 2012, pp. 11–12). It is not limited to the digital and electronic devices driving the high-tech arms race in schools and society today, sometimes confused with "innovation" and "improved learning." The science isn't in the iPad or the digital simulation. Teachers know this.

Having said that, we do need to qualify this with a word about robotics. Since we published the first edition of *Doing Good Science* in 2004, perhaps the most significant development in science education—particularly with regard to engineering—is the emergence of K–12 robotics programs in schools across the nation. In many regards, the robot embodies the role technology plays in STEM. The robot is literally the vehicle to the problem's solution—a tool, certainly, but one that requires a creative, curious, thoughtful designer. We are grateful for the explosive interest in robotics and the manner in which it has so captivated children in schools around the world; this movement is popularizing and promoting STEM and engineering design on a scale unimagined since Rubik's Cube flashed in and out of sight 30 years ago.

And we also firmly believe good science—and STEM and engineering design—is entirely possible without a single robot, or colored cube, on campus.

STEAM and STEM-D: The Role of Art and Design

It's hard to deny the irresistibility of certain technologies today, particularly given the enormous demand for them across the spectrum of countries, cultures, and commercial habits worldwide. Clearly, Apple products have captivated customers of all ages and avocations around the globe; and with due respect to the magnificent elegance of the code that defines Apple's functionality, without the artistry of their product's alluring and user friendly *design*, would Apple be (as of this writing) the 24th largest economy in the world? We suspect not.

Artists and designers are integral to STEM as the sculptors of STEM solutions that make them useful, accessible, and appealing. Arguably, scientists and artists share a penchant for creative thinking; as Max Planck, the German theoretical physicist and Nobel Laureate who originated quantum theory, asserted: "The scientist needs an artistically creative imagination." This is supported by research linking measures of increased creativity in middle schoolers with higher levels of science and math engagement (INDEX 2013).

So, we return to the *NGSS* and its use of the phrase "engineering design." As we seek to integrate the disciplines that constitute STEM, and which foster 21st-century skills like creativity, our brothers and sisters in the creative arts and humanities make a persuasive case for expanding the acronym to STEAM: science, technology, engineering, art, and math; or STEM-D, where *D* stands for "design." We agree with the argument! We hope at least one of the acronyms sticks, and requires us to do a third edition soon.

In many ways, good science has long been the crossroads of curriculum. More than any other core academic subject, science routinely incorporates key goals in literacy and mathematical reasoning, facilitated by a range of technologies, in addition to cultivating a procedural approach and higher-order problem-solving skills endemic to science and engineering practices. Despite the difficulty we might have defining it in a tidy and universally-accepted way, we believe STEM is synonymous with "doing good science."

Science and Literacy

Veteran science educators recognize the connections between science, reading, and writing. Not only are many conceptual skills transferable between literacy and science (e.g., predicting, identifying cause and effect, and using evidence), but reading and writing are also integral to good science instruction through science notebooks, lab reports, research projects, group presentations, and other elements of instruction that reflect national and state standards for language arts. Middle school science teachers can collaborate with their colleagues in other fields to identify reading, writing, speaking, and listening activities that can be conducted across the curriculum. Indeed, the broad range of engaging science topics gives purposeful writing ideas for the English teacher and support the *Common Core State Standards, ELA*'s emphasis on nonfiction subject matter.

As with classroom management expectations (see Chapter 5), our experience tells us that efforts to have an impact on student outcomes in middle school are most likely to succeed if they are implemented schoolwide. That is because middle school learners have a developmental need for structure and will respond best to instructional approaches that they encounter in a variety of classes. Furthermore, collaboration between faculty members and departments builds cohesion and purpose, supporting student outcomes.

This is especially true in schools where achievement test data can be used to identify performance patterns in individual students—for example, test results indicating which mode of writing (research, narrative, expository, persuasive, or creative) is a student's weak area. Science teachers can often be part of a team effort to improve student performance by adapting assignments accordingly. Furthermore, schools that adopt a template used consistently by teachers in every subject to teach the writing process can reinforce key skills such as presentation of main ideas, organization, voice, word choice, sentence fluency, syntax, writing conventions, and effective presentation.

Cross-disciplinary research collaborations are common in middle school between teachers in humanities fields such as English and social studies. Especially with the *NGSS* emphasis on the history and nature of science, cross-subject research projects integrating science offer robust opportunities for students to practice writing skills and support schoolwide literacy goals. The shift in the *Common Core State Standards, ELA* to a greater emphasis on nonfiction literature also opens new opportunities for science integration.

We recommend that research and writing projects in middle school science courses incorporate a writing process (i.e., prewriting, drafting, revising, editing, and publishing) and strategies that are consistent with the students' other subjects. Narratives, lab reports, essays, biographies, and even creative writing assignments based on science topics can be used to advance a school's literacy goals in attaining *Common Core State Standards, ELA*.

Science Lab Notebooks

Whenever students are involved in science and engineering practices, whether in a laboratory setting or while conducting field research, they should have the opportunity (and be expected) to record data as they progress through the process. Teachers can help students develop this essential habit by providing them with commercially prepared lab worksheets or simply giving them blank sheets of paper inserted in binders. We've found that middle-level students generate more original thoughts and observations when they start with a blank page than when they fill out a structured worksheet, although teachers will need to provide some degree of introductory preparation, especially for younger students, about what sorts of thoughts and observations are to be recorded on the blank pages.

We're pleased that since our first edition was published, science lab notebooks have become more prevalent in middle school science classrooms across the country. Students use them to record lab observations, describe findings, list questions or problems, practice journaling skills, and engage in critical reflection. Teachers may incorporate prompts into the lab activity that bring students back to the journals to think, write, and reflect. Notebooks also give the teacher a means of ongoing formative assessment to the extent that instruction is adjusted according to the needs and deficiencies identified in the notebooks.

Science lab notebooks are different from the personal journals frequently used in language arts classes, in part because they are more structured. A typical lab book activity begins with a question to investigate or an engineering problem framed by the teacher. Students then add a prediction of outcome, observations made, procedures used or models developed, and a conclusion with findings based on evidence. (See Appendix C, p. 227, "Sample Lab Report Form.") Like journals, though, lab notebook entries often involve reflection, and we know that observations and investigations lead to more questions. Teachers encourage students to present critical analyses and evidence-based argumentation in their lab notebooks based on students' experiences with hands-on activities. That is why it's important for teachers to provide time for students to write in their lab notebooks after completing investigations. Well-organized writing takes time … like good science!

We've found that teachers shouldn't assume that students automatically understand the purpose of lab notebooks or what's expected when writing in the notebooks. Students need to know how to use the notebooks and why scientists use them, including the value they hold for scientists whose work can lead to important discoveries, closely guarded secret findings, and sometimes patented solutions to problems.

We suggest modeling the entire notebooking process with the class, step by step, and displaying the sample notebook in the classroom as a visual reminder and reference for students to consult until they get the hang of notebooking. Some teachers make photocopies or circulate electronic files of the sample for students to keep in their notebooks or on their laptops as a reminder of the steps. We also find that it's helpful for students if teachers delineate expectations—where "notes" are appropriate and where to use complete sentences and pay attention to mechanics.

Students will develop ever-stronger notebooking skills through practice, especially if the teacher regularly views the notebooks, makes comments, and assigns a formal grade for completion and accuracy. (If teachers choose to formally evaluate notebooks, students should not have to guess at what it takes to get a good grade. We recommend that initially the notebooking process should be modeled and monitored for students so the expectations are clear. A checklist for evaluating the notebooks is found in Figure 4.5.)

Lab notebooks are your portal to the thinking of your budding scientists and central to the dialogue and student work in collaborative table groups (see Chapter 2), too; notebooks can help you determine how much of your instruction and influence is getting through.

Another asset of lab notebooks is that they can be infinitely manipulated to involve varied writing proficiencies, differentiated according to student skill levels, and directed to meet specific ELA standards.

Personalizing Literacy in Science

There are numerous ways teachers can make "scientific" writing more stimulating and creative for students. For example, when addressing ecologically sound ways to improve the environment, you might ask students to

- prepare a video ad convincing a neighborhood curmudgeon to recycle,
- write a letter to a legislator urging support of a clean air (or water) bill,
- outline a plan for cleaning up a vacant lot or park,
- create a recycling jingle for radio to raise the level of listeners' concern, or
- create a 30-second "drought awareness" podcast for the class or school website.

And this is just the beginning. Rick Wormeli, whose 2001 book *Meet Me in the Middle* remains a pivotal and timeless resource for teaching in the middle grades, notes that "science has many natural uses for writing, from lab reports to poetry. The blend of personal discovery and science that we might see in *National Geographic* or *Discover* magazines is achievable in our middle school classrooms" (p. 132). Wormeli recommends the following writing activities as appropriate and exciting options for students in middle school science courses:

- Write the life story of a scientist.
- Make a schedule.
- Make up a tongue twister.
- Write instructions (procedures).
- Write a consumer's guide.
- Write an origin myth.
- Create a calendar in which the picture for each month shows a particular aspect of a scientific topic.
- Write a science fiction story.
- Examine a common scientific misconception, how it is perpetuated, and what can be done to correct it.
- Explain why another student obtained certain lab results.
- Create a board game focusing on the basic steps of a science cycle or principle.
- Research and write a report about a scientific discovery that changed the world. (p. 132)

We recommend that teachers also consult with their language arts colleagues for topic ideas and possibly sharing assignments or projects.

Figure 4.4. Science Lab Notebook Checklist

The following checklist can be used to assess how well a student is keeping records during science activities. You may consider giving students a copy of the checklist to enable them to monitor themselves through the process.

General Information

- Question or problem being investigated is stated.
- All entries are dated.
- Writing conventions (e.g., punctuation, capitalization) are correct.
- Presentation is clear.

Observations

- Description is very detailed.
- Description is complete.

Illustrations

- Drawings are accurate.
- Drawings are in color
- Drawings are labeled.

Procedure

- List materials used.
- Sequence steps followed.
- Communication of data
- Graphic/table is complete.
- Graphic/table is labeled.
- Graphic/table is mathematically correct.
- Written portions are clear and complete.

Analysis, Argumentation, and Conclusion

- Analysis is clear and logical.
- Explanations and claims are made using evidence from the data.
- New questions, models, or investigations are proposed.

Line of Learning

- New learning is shared.
- New curiosities are shared.

One of our favorite writing prompts after a lab or activity—which is deceptively simple and yet engages higher-order skills—is "Explain to Ellen." Ellen was absent yesterday; explain the lab to her, what discoveries were made, problems solved (or encountered), what questions emerged, whether the investigation was a success or not and why, and so forth.

A fun departure that can foster collaboration with social studies colleagues invites students to research a scientist and take on that individual's persona, dressing in period costumes and speaking in first person about their work in the lab or doing field research. This can be challenging—they can't say "I was born on October 5, 1900 and died on May 3, 1974" because dead people can't talk! But they still need to convey all the key facts of the scientist's life. Truly, the possibilities are limited only by a good science teacher's imagination and ability to inspire and productively channel the creative energies of students!

Reading and Science

In addition to the many writing opportunities that good science instruction makes possible, it also helps students develop reading comprehension skills. Teachers can use the following questions from Thier and Daviss (2002) to help students learn to reflect on science writings and develop their reading and research strategies:

Reading Comprehension Prompts for Students

- Predicting
  - With a title like this, what is this reading probably about?
  - What will happen next? (Turn to your partner and tell what might happen.)
- Reflective questioning before reading
  - Why am I reading this?
  - Why does the author think I should read this?
  - What do I expect to learn from reading this?
  - How does this relate to my life?
  - What do I already know about this topic?
- Reflective questioning after reading
  - What do I still not understand?
  - What do I still want to know?
  - What questions do I still have about this topic?
- Paraphrasing or retelling
  - What was the reading about?

- Can I explain to my partner or group, in my own words, the meaning of what I just read?
  - Summarizing
    - Can I identify all the key concepts from the reading and write a summary using these concepts? (pp. 42–43)

Other prompts we've used with good outcomes are, "What is the author's purpose in writing this?" and "How does it relate to me? Or to life today for everyone?"

Good science instruction also incorporates speaking and listening activities through presentations, projects, discussions, and reports, which are other areas of skill development that support the *Common Core State Standards, ELA*.

Science Is Mathematics

Or put another way, math is a language for science. Like writing and reading, math is an essential part of good science instruction, and *can't be separated from it* because it's woven inextricably into activities and experiments.

While the two disciplines are widely perceived as inseparable, however, at times we may miss the opportunity to reinforce mathematical concepts, skills, or modeling in a way that will enhance student proficiency and achievement. Math concepts are present in scientific operations such as graphing, predicting, measuring, weighing, and collecting and analyzing data, but teachers must help students see the connections between the mathematical processes that are embedded in science activities and the mathematical principles students are learning in their math classes.

Especially considering the emphasis in *NGSS* on performance-based assessment, teachers are called to expect that students show the math calculations used to determine their results. They will soon get the picture—"Wow! We are doing math!"—and that is a powerful teachable moment.

Math is abundant, if embedded, in good science. Beyond basic computation, well-planned good science activities typically call for students to engage in estimation, proportionality, and even basic algebraic and geometric concepts—topics identified as weaknesses in U.S. students' science performance in international comparisons. Middle school science teachers should focus on developing activities that explicitly engage students in these mathematical operations, especially algebra and geometry operations. Please refer to the activities to see how mathematical processes and the *Common Core State Standards, Mathematics* can be incorporated in good science activities.

Math and science are sometimes lumped together as "aptitudes" that some students have and others don't. In math as well as science, the middle grades are where students begin to develop an identity as a learner—successful or unsuccessful—in these subjects. As in science, many of our colleagues involved in math coordination and training believe that mechanical, teacher-centered methods are one reason middle schoolers perceive themselves as "bad at math." We believe that the integration of math into good science instruction must adhere to the same instructional principles that we discuss throughout this book.

Good Science Can Be Low Tech

As we mentioned earlier in this chapter, it is important to distinguish between the common use of the term *technology*—i.e., digital media, probeware, and the like—and its meaning as defined in the *Framework*, examined earlier in this chapter, as it relates to doing good science: "Technologies result when engineers apply their understanding of the natural world and of human behavior to design ways to satisfy human needs and wants" (NRC 2012, pp. 11–12).

Technology should be used to meet a need or solve a problem, but this doesn't mean you require digital media and expensive electronic probeware to conduct activities featuring science and engineering practices. Moyer and Everett point out that "every day we use common, ordinary objects, such as scissors, can openers, and zippers, that involve quite sophisticated engineering" (2012, p. 9). Technology can simply be the means to an engineering design solution and can involve little more complexity than challenging students to design and build shockproof containers used in an egg drop or to recombine everyday materials to invent a better mousetrap—without a mousetrap.

We discount the need perceived by some teachers and schools to purchase sophisticated digital technology programs or specialized hardware for science instruction. The technology is just a tool to use during an investigation or an engineering design step. When we do science, sometimes we have all the tools we need and sometimes we don't. Good science is not dependent on hardware and software. Good science can be low tech.

That's not to say that we're opposed to using scientific technology, if a school or district can afford it. Indeed, we've worked with fifth graders in scaled-down electronic flight simulators to learn the physics of flight and with eighth graders using night-vision goggle technology to study rock formations under desert starlight. We encourage teachers to pursue technology resources and training to use technology, as many investigations can be enhanced by technology. We also appreciate the advantages of using prepackaged science kits, particularly when they are part of a systemic plan for doing good science and accompanied by training from the kit vendor. However, the absence of science kits or digital media does not prohibit teachers from implementing science and engineering practices or from providing students with exciting learning opportunities.

References

Brunsell, E., ed. 2012. *Integrating engineering and science in your classroom.* Arlington, VA: NSTA Press.

Gerlach, J. 2012a. Elementary design challenges: Students emulate NASA engineers. In *Integrating engineering and science in your classroom*, ed. E. Brunsell, 43–47. Arlington, VA: NSTA Press.

Gerlach, J. 2012b. STEM: Defying a simple definition. *NSTA Reports,* April 11. *www.nsta.org/ publications/news/story.aspx?id=59305*

Independent School Data Exchange (INDEX). 2013. Proceedings of the annual meeting of the Mission Skills Assessment (MSA) schools, Chicago.

Moyer, R., and S. Everett. 2012. *Everyday engineering: Putting the E in STEM teaching and learning.* Arlington, VA: NSTA Press.

National Academies. 2007. *Rising above the gathering storm: Energizing and employing America for a brighter economic future.* Washington, DC: National Academies Press.

National Academies. 2011. *Successful K–12 STEM education.* Washington, DC: National Academies Press.

National Research Council (NRC). 2012. *A framework for K–12 science education: Practices, crosscutting concepts, and core ideas.* Washington, DC: National Academies Press.

NGSS Lead States. 2013. *Next Generation Science Standards: For states, by states.* Washington, DC: National Academies Press. *www.nexgenscience.org/ next-generation-science-standards.*

Osborne, J., et al. 2012. Educating students about careers in science: Why it matters. In *Exemplary science for building interest in STEM careers*, ed. R. Yager, 51–62. Arlington, VA: NSTA Press.

Thier, M., and B. Daviss. 2002. *The new science literacy: Using language skills to help students learn science.* Portsmouth, NH: Heinemann.

Vasquez, J., C. Sneider, and M. Comer. 2013. *STEM lesson essentials, grades 3-8: Integrating science, technology, engineering, and mathematics.* Portsmouth, NH: Heinemann.

Wojnowski, B., K. Charles, and T. Warnock. 2012. Why STEM? Why now? The challenge for U.S. education to promote STEM careers. In *Exemplary science for building interest in STEM careers*, ed. R. Yager, 63–80. Arlington, VA: NSTA Press.

Wormeli, R. 2001. *Meet me in the middle: Becoming an accomplished middle-level teacher.* Portland, ME: Stenhouse.

Yager, R., ed. 2012. *Exemplary science for building interest in STEM careers.* Arlington, VA: NSTA Press.

Classroom Management and Safety

Welcome to the challenge of making good science come to life in your classroom. In this chapter, we look at how to get the classroom ready for good science lessons and how to prepare students so their activities are engaging, productive, and safe.

The essence of classroom management in the middle school is teaching students to expect and follow *procedures*. If middle grades teachers accomplish that task early, and reinforce it throughout the term, they can expect significantly fewer surprises and hassles when it comes to student behavior, lab safety, and upkeep of equipment and materials.

The First Days With Students

> What you do on the first days of school will determine your success or failure for the rest of the school year. You either win or lose your class on the first days of school. (Wong and Wong 2006, p. 3)

Harry Wong and Rosemary Wong's assertion from their classic book for new teachers appears bold, but it is immediately relevant to teaching in the middle grades. Teachers preparing for the first days of middle school science need to tackle the following crucial issues before students ever enter the classroom door:

- classroom arrangements that enable success (e.g., strategic groupings of students in collaborative table groups, as we discuss in Chapter 2; posting classroom procedures and safety expectations; publishing exemplary student work around the classroom, etc.);

- an idea of the instructional "big picture"—that is, where you want to go, in keeping with the *NGSS* and your school's curricular program; and

- lessons and assessments that bring about continuous instructional improvement and student achievement.

We should add that we don't recommend designing your entire curriculum over the summer. Good science is most often achieved through small steps and "subtle shifts," as San Francisco's Exploratorium calls them, guided by an overall plan. The key concept here is recognizing that much of the accomplishment attributed to master middle school science teachers stems from their preparations for the opening of school, which, in turn, are founded on their expectations for themselves and for their students.

In a learner-centered classroom, the teacher must be aware of the range of needs her students will bring to class on those first days. Middle schoolers start the school year in varying degrees of preparedness to learn, according to their home and family situations, health and nutrition, attitudes toward teachers and schooling, and previous experiences with science as a subject. In the special case of students who are moving to middle school for the first time—where they will switch to a departmentalized class schedule with five or more teachers, bell schedules, and passing time between classes—a number of basic uncertainties arise: Who will my teacher be? Will she be nice? How hard is science in middle school? Is there a lot of homework? Will I make friends in the class? And so forth. The answers to these questions, among others with which students begin the year, shape student expectations, attitudes, and behaviors. And all of their questions must be carefully considered due to the safety concerns, student dynamics, and the interactive nature of good science in the middle grades.

Classroom Management

Management vs. Discipline

This is the section in which we connect the dots, so to speak, between the traits and needs of the middle school student described in Chapter 1, the characteristics of good science instruction discussed in Chapter 3, and the procedural strategies teachers use to successfully channel bursts of adolescent energy in a meaningful direction. It can be done!

At the outset, we need to make a distinction between *discipline* and *classroom management.* In our experience, *discipline* almost always entails reacting to disruptions and then assigning consequences, usually negative; *classroom management* is about planning and preparing procedures to maximize student time-on-task and productivity. The most effective middle-grade teachers structure their lessons—from the pivotal first few days of class—using procedures that students learn and internalize as routine.

This is not to say a teacher shouldn't have rules, which are what most teachers associate with "discipline"; indeed, a limited number of rules are an integral part of the classroom management plan. A middle school teacher who is a "disciplinarian" tends to post a list of rules—or sometimes may involve students in writing the rules and then post them—as an accountability measure: "The rule is posted up there [teacher points with an index finger]; therefore, if you break it, you knew better and deserve what you get." In this sense, the rules are about deterrence—punishment—and are not integrated into the learning continuum.

An effective classroom manager may post the same rules—literally—and may or may not involve students in determining what they should be. The key difference is that a classroom manager teaches the rules to the students over the course of the first week or so of school, in conjunction with other procedures for taking attendance, training students to begin each class by sitting down and working on a short assignment posted on the board, breaking down labs with 5–10 minutes remaining in class, and being dismissed (where appropriate) only when the teacher says so. "Discipline"—or fear of punishment—now becomes "classroom management"—a learning process that manifests as a practiced behavior. This is the power of teaching procedures.

Perhaps the cardinal rule governing the behavior and attitudes of middle schoolers in good science classrooms would be something like this:

> As long as you act like a scientist when we do science, you will be treated like a scientist and enjoy many exciting investigations and discoveries. If not, you get to watch the rest of us do science! (Variation: You get to be a retired scientist!)

This expectation has served us (and many of the teachers we consulted in writing this book) as very effective management messaging with middle-level students.

In Table 5.1 (p. 60), we list a sampling of ineffective methods and effective procedures that illustrate the advantages of a well-planned approach to classroom management.

Procedures, and rules in particular, satisfy a pre-adolescent's need for structure, safety, and predictability, as discussed in Chapter 1. Part of the objective when teaching about the rules is to make students aware of the need to follow procedures to avoid unfavorable outcomes, such as accidents. We're not suggesting that classroom management plans not feature consequences for poor choices. Indeed, the most effective plans include rules with consequences that logically fit the student behavior in question. Middle school students need, and on a certain level *want* to know the limits, but they are also looking for evidence that the world and adults are basically fair.

That is why we advocate "teaching the rules" rather than simply posting them like a skull and crossbones! Serious infractions and disruptions, and certainly any behavior that endangers another individual in the classroom, demand major but also reasonable consequences, and middle school students should be taught why the consequences follow—and fit—misbehavior in the science classroom.

Table 5.1. Ineffective Methods vs. Effective Procedures for Middle School Classroom Management

Ineffective Methods	Effective Procedures
Let students sit where they like.	Assign student seating within first week.
Post rules on the wall; refer to them when students act inappropriately.	Teach and test classroom expectations in first weeks; post and reinforce throughout year.
Leave parents out of the loop until a situation escalates, course grade drops, and/or end of term nears.	Involve parents in classroom management with early positive communication and follow-up.
Neglect to properly educate students what to do in the event of an absence, an emergency, a substitute teacher, a missed assignment, or unexpected "free" time.	Establish class procedures well in advance about absences, emergencies, and unexpected "free" time.
Take attendance while students chat/sit idly.	Start class with a brief student task.
Begin teaching without outlining the expectations or objectives	Post daily schedule, including lesson, start-and end-of-class activities (bell work), what's due, activity/lab, and task(s).
Announce assignments verbally, with little notice; no student accountability for recording assignments or completing or submitting them.	Post assignments in advance to be recorded and checked in by students on calendars they maintain (which are checked periodically by the teacher).
Correct inappropriate work habits as they occur; limit cooperative/lab activities based on poor behavior.	Teach students how to work independently and in groups, how to get your attention, and how to clean up after a lab or activity.
Respond inconsistently to requests to leave the classroom based on circumstances.	Train students how to move about or leave the classroom during seatwork and labs.
Respond to inappropriate questions or comments.	Teach students how to offer and respond to questions and criticism.
Give free time at end of class (e.g., allow students to congregate at door).	Instruct students how to end a class and let them know that you, and not the bell, are in charge of when they can leave.

In the best-managed middle school classrooms we've seen, rules, consequences, and management procedures are *taught*—and tested (or quizzed)—during the first week of school and throughout the year. (Once they're taught in that first week, we recommend they be posted as reminders and for reference.)

Some effective middle school teachers prefer to determine and post the rules without consulting their students; others use a more democratic approach. Either approach can be compatible with classroom management plans tailored to adolescent learners and to science and engineering practices, although the more-participatory systems help to develop a sense of classroom community that is common in the effective middle school classrooms we've known.

Beyond developing classroom unity, successful middle school educators strive to develop broader agreement when it comes to classroom management by collaborating with other staff to adopt schoolwide behavior expectations—for students *and* teachers.

A Strategy to Avoid

There is a "discipline" strategy teachers sometimes employ that we strongly discourage. In keeping with the goal of building community (and in the spirit of developing relationships with students), we believe teachers should avoid publicly singling out or humiliating individual students for misbehavior.

A better way can be as simple as asking a student to step outside briefly to avoid a public confrontation or to meet for a discussion at lunch, homeroom, or recess, depending on the circumstances of the situation and the need for immediate intervention. (A private conversation about expectations eliminates many peer pressure complications.) Often a calm, respectful request to confer one on one is all it takes to encourage a student to refocus his or her attention, and the lesson can continue with minimal disruption.

Other times, it's not quite that easy—but remember the teacher's objective should always be to build relationships and community rather than break down a student's will or self-esteem. This should be a guiding principle of a middle school teacher's classroom management efforts. Presence of mind, objectivity, and several deep breaths will pay off in the long run.

Rewards and Praise

Negative consequences are commonplace middle school disciplinary approaches. However, many middle-level educators also incorporate positive incentives such as praise and rewards into their classroom management plans. Is positive reinforcement an effective management strategy? The answer is, it depends. Some experts argue vehemently against the use of tangible rewards such as candy and stickers. We are conditioning our children to expect a reward for behaving appropriately, the argument goes, and reinforcing their need for extrinsic motivation while diminishing their intrinsic drive. Teachers who implement a reward system will have profound difficulty eliminating it later in the year and may find students expecting rewards for an ever-expanding array of "normal" behaviors. Meanwhile, teachers become so dependent on doling out rewards as incentives that they may be less aware or inclined toward other, more creative and effective motivational strategies.

That said, some forms of rewards are effective and appropriate in the middle grades. Our experience has been that specific positive reinforcement in the form of verbal praise is just as effective (and healthier) than tangible rewards such as the ubiquitous candy handouts. Praise applied appropriately, specifically, and sparingly appears to have a positive impact on student attitudes and behaviors when it is contingent on students having reached a set standard or expectation.

Acknowledging Differences

Also in regard to classroom management, we urge teachers to be sensitive to the behavior patterns of children from cultural or demographic backgrounds that are different from their own. When teachers fail to take into account such differences, they can develop management procedures that are ineffective or that provoke a very different reaction than the teacher intends.

Are we advocating different rules for different students or that teachers apply consequences inconsistently? Perhaps this question is best addressed by asking another question: Is the goal of classroom management in the middle school to achieve "equal" treatment of all students? Our answer is no, because, as we discussed in Chapter 1, an important emotional need of middle-level learners is to be treated as individuals, with the teacher showing sensitivity toward the specific circumstances and emerging identity of each student.

Just as we differentiate to address the unique learning needs and abilities of children, we're also called to consider the behavioral profiles, backgrounds, and personas of individual middle schoolers so we can use the management tactics that will work best for each class combination. In fact, when it comes to disadvantaged students—who as adults are profoundly underrepresented among the ranks of science teachers, scientists, and scientific professionals because proportionally very few of them attend college—we are eager to discover and pursue new ways that will help all students succeed. In this regard, we direct readers to the important work of Ruby Payne (2005), whose training for educators on strategies for teaching children living in generational poverty is extremely useful for at-risk school communities in general. The future of good science—inclusive of diverse backgrounds and experiences—depends on this sort of resourcefulness.

Lesson Planning for Good Management

Basic to good science teaching is establishing an atmosphere of "investigation fervor." Early in the year, teachers need to let students know that they *are* scientists and that in the course of the term they will use and reuse their skills in the class. What skills are necessary for students to think and act like scientists? First, teachers need to make students understand that scientific research always starts with a question, a puzzling observation that raises a question, or in the case of engineering activity, a problem to be framed and addressed. This is a platform for introducing (or reintroducing) students to processes they can use to conduct investigations (discussed in more detail in Chapter 6).

Second, teachers must introduce students to important science skills, including

- observing and collecting data,
- estimating,
- framing problems and designing solutions,
- identifying variables,
- predicting, hypothesizing,
- investigating and asking questions,
- measuring,
- classifying,
- building and testing models,
- making graphs,
- discovering or determining cause and effect,
- making inferences,
- communicating,
- drawing conclusions, and
- supporting conclusions using evidence.

Teachers should design lessons to activate these skills while helping students become more comfortable asking questions, defending or justifying their answers, and generally being skeptical of results and conclusions. These process skills are generic and applicable to any topic a teacher may choose to teach within the scope of the *NGSS*.

Providing students with lab settings that maximize use of the greatest number of these process skills will expedite their competency and assist them in thinking like scientists.

One teacher we know makes a habit of leading students to identify the science skills they've used in a lab activity or investigation. Each time a new skill is mentioned, the students create and decorate a placard or banner to post on the classroom wall, forming an impressive (and creative) visual record of the many skills the students are acquiring by doing good science.

Teaching Safety

Middle schoolers are by nature incredibly curious young people. This trait, combined with their tendency to explore the limits of appropriate behavior, forms a potent recipe for science disasters. Consequently, no discussion of classroom management in the context of middle-level science education is complete without considering the issue of student and teacher safety. Safety is primary to good science—everything else is secondary.

Indeed, we strongly recommend that teachers consider doing a safety refresher training, perhaps a webinar, available online through various vendors and associations; see also NSTA's "Safety in the Science Classroom" (*www.nsta.org/pdfs/SafetyInTheScienceClassroom. pdf*). Increasingly, larger districts appear to be implementing science safety training to protect teachers from liability claims. What limited training most of us received in our teacher preparation coursework is likely outdated and not well aligned with the degree of activity associated with science and engineering practices.

Hands-on, minds-on science experiences present more risks than those associated with reading textbooks and completing worksheets. The professional judgment of the teacher based on legal safety standards and professional best safety practices is the most important factor in determining which activities should be used and which activities should be omitted. Accordingly, science safety is maximized by teachers who set clear expectations for student behavior.

As with all management expectations, safety is a matter of teaching, testing, and reinforcing procedures. Questions about safety, particularly issues surrounding lab activities, should appear in the start-of-term student pre-assessment to determine the extent of the students' previous safety preparation.

In addition, prior to engaging in any science activities, students must be required to complete and return a safety acknowledgement form signed by their parents and also score 100% on a safety quiz. (Readers may refer to the various science materials vendors listed in Chapter 18 to find sample acknowledgement forms and quizzes.) Safety acknowledgement forms advise students and parents that the lab can be an unsafe place. To make it safer, safety procedures listed in the acknowledgement form must be followed. We urge teachers to include a specific clause in the acknowledgement form that warns against unauthorized experimentation—good science is fun, but teachers are professionals and students should *not* try this at home!

Key procedures that we recommend teaching and implementing from the start of the term are the following:

- Make it clear that safety is the highest priority in your classroom and that students who choose not to behave like scientists during an activity or project will observe, rather than participate, for that day.
- Familiarize and train students regarding location and use of sinks, eyewash, chemical shower, first-aid kit, emergency shutoff switches, and other emergency stations.
- Model appropriate use of personal protective equipment such as indirectly vented chemical splash goggles, safety glasses, gloves, and aprons. Also show students how to sanitize goggles and glasses.
- Model correct use of equipment such as heat sources, hand tools, lab ware, and so on.
- Assign a materials manager for each group who inventories items before and after the activity.
- Designate a signal to indicate cleanup and inventory time.
- Establish an expectation for cleanup and materials inventories so that students understand class won't be dismissed until the room is in order and all supplies are accounted for.
- Make sure that materials are organized and packaged in a box or tub for each group.
- Locate materials so students don't have to carry them too far (how far is too far depends on the hazards associated with what is carried). Have materials well labeled in the classroom for ready access.
- Arrange the physical space in the classroom to accommodate the traffic patterns of the specific lab activity, including access to safety equipment, as well as setup and cleanup considerations.
- "Declutter" prior to lab days, as clutter compounds the possibility of accidents and injuries.

There Is No Substitute for Safety!

1. Train your students in safety and emergency procedures.
2. Provide equipment in good operating condition. Take equipment out of service immediately if damaged.
3. Supervise students AT ALL TIMES!

A lab safety checklist, which can be posted in the room following the teaching of a safety unit at the start of the year, might look like Figure 5.1. (See also the lab safety rules in Appendix E, p. 233.) Teachers should review every item, with every student, before every lab.

We live in an increasingly litigious society, and these sorts of steps can help protect teachers and their school districts in the event of an accident that leads to litigation. We know of some teachers who ask students and parents to sign a "lab safety contract," at least in part to mitigate their liability. However, as Kenneth Roy (2012) points out, "Safety 'contracts' are not legally binding when signed by students of middle school age. They should be referred to as 'Safety Acknowledgment Forms' that can protect the teacher should an accident occur and litigation were to follow. The teacher is fostering 'duty of care' by informing parents of the inherent dangers in a science lab and asking for their support in working with students" (p. 131).

Proper hand-washing is a fundamental routine to be taught and practiced in the good science classroom. The Centers for Disease Control recommends washing with soap and running water for 20 seconds—"[the equivalent of] singing Happy Birthday twice! The CDC also states that if soap and water or not available, alcohol-based hand-sanitizer products should be used (60% ethanol)" (Roy 2012, p. 64).

We celebrate curiosity and are excited when our students are interested enough to pursue their investigations. Increasingly, however, very dangerous "science" activities are available to young people with access to the internet and other sources. Years ago, this prompted the "See Me" rule, which is as follows:

> If you want to try your own experiment, please SEE ME first! We can investigate it together, and I will evaluate the safety of the proposed experiment—bring instructions or information if you have any. If it is safe, you may proceed with the investigation under my direct supervision.

After the first few times we explained the rule, we'd ask in a lab write-up or quiz for students to repeat it. This helped emphasize the importance of open communication between teacher and students and reiterated the gravity of the expectation.

We also made sure the parents understood "See Me." For added safety, we recommend that teachers require all independent investigations to include a complete description of the proposed inquiry with verifiable signatures of the parent(s) indicating they understand that all procedures must take place at school under the teacher's supervision. (There have been cases where teachers were held liable for student "science experiments" at home when the teacher talked about specific subject matter in class and neglected this sort of explicit caution.) Furthermore, we advise that if a teacher is inexperienced or unfamiliar with the specific area of a student's independent study interest, it's essential to seek the advice of experts to help evaluate the safety and educational value of the student's proposal.

Figure 5.I. Lab Safety Checklist

Before You Begin

1. Do you know who your lab team members are?
2. Do you know the task for which your lab team is responsible today?
3. Do you know the procedures for teamwork?
4. Do you know proper safety precautions and personal protective equipment needed for today's lab (goggles, apron, gloves, etc.)?
5. Do you know who to contact in case of a classroom emergency?
6. Do you know where basic safety supplies (baking soda, eyewash, water) are kept in case you are asked to locate them?
7. What is the *first* thing to do in the event of an emergency?

After the Lab Is Completed

1. Did your team work well together?
2. Did your team complete the assigned task?
3. Did your team follow all the safety rules?
4. Did your team need to use any first-aid supplies?
5. If yes, which ones and why? (Remember to restock!)

We've discovered another hazard in the good science classroom that didn't appear to be an issue when we wrote the first edition of this book a decade ago: the prevalence of acrylic nails among our students. The U.S. Food and Drug Administration warns that some acrylic nails can easily catch fire: "For safety's sake [when lighting matches or Bunsen burners], either ban acrylic nails or require appropriate glove protection when working near flames in the laboratory" (Roy 2012, p.129).

Transitioning to a Green Science Classroom

Increasingly, we're aware that exposure to toxic chemicals and substances should be minimized as much as possible, and certainly more so than was allowed in the past. Some science teachers we know have revisited their position over the years and now advocate moving toward the goal that "nothing but water goes down the drain in this classroom." This is not meant to prevent good science but to encourage responsible and safe chemical disposal and exploring new alternatives about old ways of doing good science. Roy recommends that middle school science teachers "work with their custodial departments to transition into using greener cleaning products in their rooms. A well-designed green cleaning program can help reduce harmful exposures and yield many other benefits for students, custodial staff, administrators, and the environment" (p. 75). In all cases, we urge teachers to be mindful of the chemicals and substances they use and to ensure that all wastes are collected and disposed of properly.

On a related note, Roy encourages heightened sensitivity to the impact of invasive species that might be derived from improper disposals or discards. He urges middle school science teachers to "plan ahead and determine what types of live plants and animals provide for a good instructional learning experience and also do not affect the local environment upon release" (2012, p. 118).

An Emotionally Safe Environment

Beyond sensitivity to ecologically sound products and practices and a conscious concern for safety in the classroom, the environment should be tailored to the needs and interests of students and teachers who have to live together in the room for months. We've seen a wide variety of middle school classroom environments that suited student needs and were associated with successful students and teachers year after year. Generally speaking, they were places with some color, warmth, and public acknowledgment that students are important to the teacher (e.g., student work posted, birthdays listed). Some teachers find that creative and varied desk arrangements facilitate different instructional activities better (and can cause a fairly marked difference in how students behave and interact). In Chapter 2, we saw that middle schoolers need a blend of spontaneity and structure in a teacher's instructional methods; this is probably a good rule of thumb concerning the class environment, too.

In sum, we agree with Rick Wormeli's tried-and-true call for creating "an emotionally safe environment" (2001, p. 8), focusing on a teacher's role of empowering learners in the middle grades:

> There are many ways to boost the confidence levels in our middle school classrooms without getting lost in self-esteem hoopla such as putting up "happy" posters. Be pleasant to students. Call them by their first names. Greet them at the door. Smile often. Catch them doing something well. Crack a few jokes. Ask questions that show your interest. Applaud risk taking. Share excellent homework or test responses with the rest of the class. Allow occasional democratic voting in the class. Refer one child who is an expert on something to another child who needs help, and make sure you rotate the expert's role. Ask students to tutor their peers after school. Give them responsible jobs in the classroom. Ask them to serve as hosts for guest lecturers. Point out moments of caring among peers that occur in class. Post their accomplishments in class. Make at least one positive phone call or note home for each child per year. (pp. 8–9)

Efforts such as these will go farther, based on our experience, than any number of pop culture icon posters or slogan banners to win the loyalty and interest of middle school learners.

References

Payne, R. 2005. *A framework for understanding poverty, rev. ed.* Highlands, TX: RFT Publishing.

Roy, K., 2012. *The NSTA ready-reference guide to safer science, vol. 2.* Arlington, VA: NSTA Press.

Wong, H. K., and R. T. Wong. 2006. *The first days of school: How to be an effective teacher.* Mountain View, CA: Harry K. Wong Publications.

Wormeli, R. 2001. *Meet me in the middle: Becoming an accomplished middle-level teacher.* Portland, ME: Stenhouse.

CHAPTER 6

Good Science Activities in Action

Questioning, Framing Problems, Differentiating, and Assessing

As preparation for the 10 activities later in the book, let's look at several important considerations for carrying them out. To do that, we will address five topics: (1) using questions to guide good science investigations; (2) framing problems for engineering investigations; (3) differentiating activities to meet the range of abilities and needs in a typical class; (4) shifting from traditional teaching (direct instruction, demonstrations, and lectures) to incorporate active science and engineering practices; and (5) assessing students in a good science class.

Questioning strategies in general—and the use of focus questions in particular—are a foundation for effective middle school science instruction. That's where we'll start.

Using Focus Questions

Focus questions may be used in a variety of ways. The teacher can ask them before a demonstration or student lab activity as a form of "anticipatory set" to focus and motivate students. These questions would involve prompts like, "What do you think will happen if …?"

More often, the teacher uses focus questions while circulating around the classroom during an activity to encourage students to investigate something they have observed, have students probe an unexpected outcome or discrepant event, or explain counterintuitive findings. Teachers also use focus questions to ascertain or refresh prior knowledge, reinvigorate a stalled investigation, help students discover how they might navigate around misconceptions or errors, and evoke further questions and investigations.

The following list is a short sampling of focus questions and question stems to illustrate this strategy:

- What are you trying to demonstrate or test?
- What have you determined so far?
- How do you know that? (What evidence suggests that?)
- Why do you think that happened?
- What didn't work?
- What have you tried? Why? Could we try it another way?
- Is there anything else you could try? What else could you do?

- Is it possible to …?
- Have you considered …?
- What if you …?
- Where could you look …?
- Tell me how you …
- What other information do you still need?
- What did you discover when …?
- Do we all agree about this?
- What is going on when …?
- What do you see? Is that all you see?
- What questions do you have?
- What would happen if …?
- How might that have been different?
- What does that suggest to you?
- Is there someone you could ask?
- Looking back, tell me about …
- What is it exactly that you want to find out?
- What data do we need? How can we collect that data?

Of course, students should be encouraged to generate their own questions too, and the most capable young scientists may be able to conduct their own investigations with very little teacher influence. In our experience, however, good science in middle school is most often successful (and practical) when the teacher guides the launch of an activity with focus questions or help in framing the problem. The teacher nudges and suggests through questions and brief interviews, offering sample problems and asking students to suggest first steps; attempts at self-directed, student-led investigations can come later in middle school with more experienced youngsters.

Students new to hands-on, minds-on science will occasionally hit dead ends; they can also become confused or frustrated when their progress is slow. This is when they need their teacher's help, and focus questions are a key resource for the teacher. Exactly which questions you use or how much of the problem framing you do for the table groups will depend on the nature of the students and their investigation, but it's vital that the teacher is moving around the room as a co-investigator, prepared to pose questions at critical moments in the activity. The teacher's goal is to help students gain confidence in their ability to pursue an investigation with increasing independence.

Framing Engineering Problems

Identifying the problem is critical in engineering design, because without knowing what the problem is, we pursue the wrong questions and test the wrong solutions.

According to the *Framework*, "the engineering design process begins with the identification of a problem to solve and the specification of clear goals, or criteria, that the final product or system must meet" (NRC 2012, p. 204). This is called "framing the problem," and in middle school engineering challenges, a framed problem is generally provided to students by the teacher.

However, between these defined parameters of the initial problem and outcome criteria, students must identify the specific steps, equipment, and materials necessary to develop solutions. Next, they need to determine which solutions are most viable, and then decide the solution prototypes to build and test.

It is in this creative and generative stage of engineering design, between problem and solution, that most errors (and learning!) can take place.

A Differentiated Good Science Classroom

A fundamental goal of differentiation is to encourage teachers to be aware of and responsive to the range of needs among their students. As Carol Ann Tomlinson and Jay McTighe assert, "Responsive or differentiated teaching means a teacher is as attuned to students' varied learning needs as to the requirements of a thoughtful and well-articulated curriculum" (2006, p. 18).

Central to a differentiated, learner-centered teaching approach is the recognition that students acquire, process, and demonstrate mastery of knowledge differently. This is especially true in middle school, where youngsters experience profound developmental changes at rates that vary widely from one student to the next. It is entirely possible to have a single middle school science class with students functioning at both extremes of the cognitive spectrum—concrete thinking and complex, abstract thinking—and everywhere in between.

Teachers serving in diverse school settings know that children come to a classroom with backgrounds and experiences that profoundly shape their learning. We taught Native American children in Arizona who could not study owl pellets or snakes because of their tribal belief systems. In larger districts, students coming from different elementary feeder schools may be accustomed to different science protocols, teaching styles, and expectations.

It can take time to bring a class of students with disparate skills, backgrounds, and experiences to the threshold of good science. However, in our experience, once middle school students become engaged in their own learning—because it's active rather than passive, it's relevant to their experiences and interests, it's challenging (and even *fun*), and it affords them a degree of independence that they crave—they are both empowered and inspired to pursue further learning.

As we explored in detail in Chapter 1, it is during the middle school years when many children form their *dispositions* toward learning: "Am I good at writing?" "Do I like science?" The degree to which a teacher is responsive to the needs, interests, and abilities of students can significantly shape their perceptions, and "teachers in effectively differentiated classes help students participate in the formulation of their own identity as learners" (Tomlinson 2008, p.30).

For these reasons, learner-centered lesson planning calls on teachers to make certain accommodations that will increase the likelihood that good science teaching will take hold (Table 6.1).

Table 6.I. Traditional Classroom vs. Learner-Centered Classroom

Traditional Classroom	Learner-Centered Classroom
Emphasis on direct instruction prioritizes teacher's preference	Emphasis on differentiated instruction prioritizes student needs and learning styles
Whole-group instruction predominates	Mix of whole-group and small-group instruction including centers
Competitive ethos (emphasis on individual success)	Collaborative ethos (emphasis on teamwork)
Measure progress against a goal toward which the whole group aspires	Measure progress against where individual children started
Learning is passive	Learning is active
Low-level cognitive skills such as memorization and recall prevail	Learning activities engage the full range of cognitive skills
Teacher's interests and priorities dominate; students see limited relevance in material	Student interests play a role in instructional decisions; teacher strives to make learning relevant to students
Teacher sets uniform assignments (i.e., everyone does the same essay)	Teacher offers students choice of assignments and projects
Limited or no opportunity for creative expression or projects	Frequent opportunities for project- and problem-based learning
"One answer" mentality prevails (i.e., emphasis on multiple choice tests)	Creative approaches and argumentation encouraged in problem-solving
Summative assessment dominates or is the exclusive method used; objective is to measure learning of whole group and rank students according to their achievement	Multiple assessment forms used, with emphasis on formative assessment; objective is monitoring individual student progress and adjusting instruction accordingly

Key to effective differentiation in the middle school science classroom, Doug Llwellyn asserts, is providing students (varying degrees of) control and *choice* in their learning experience.

> Choice is an authentic means to increase intrinsic motivation, where students complete a task for its own sake and their own satisfaction. Effective teachers focus on fostering intrinsic motivation that encourages self-determined and self-fulfilled autonomous students, moving instruction from "you have to" to "you can choose to" and shifting ownership of learning from the teacher to the individual student. (2013, p. 90)

Offering choices is powerful fuel for igniting independent learning. Surveying and engaging student interests is another means of tapping intrinsic motivation and sends the message to students that a teacher is interested in what *they* think. When enlisted as a pre-assessment, a science interest survey can be a helpful tool at the beginning of a school year to begin differentiating according to student preferences, learning profiles, and need for structure. See Figure 6.1 for a sample science preferences survey.

Figure 6.1. Sample Science Preferences Survey

For each statement that follows, indicate whether you agree (A), disagree (D), or are neutral (N):

1. I enjoy learning about animals and plants.
2. I enjoy learning about energy and matter, like electricity and atomic particles.
3. I enjoy learning about the Earth and space.
4. I prefer to learn about science by reading the text and listening to the teacher rather than by doing experiments.
5. For a major grade, I'd rather build and test a model bridge than take a written exam.
6. I like working alone more than in a group.
7. I like to learn by watching others do a task first.
8. I like to learn by exploring and discovering by myself or in a group, rather than being told what to do by the teacher.
9. I enjoy presenting my projects to the class.
10. I'm good at using technology, including computers.
11. I enjoy creating artwork as part of an assignment.
12. I'm good at doing research.
13. I like to debate and prove my side of an argument.
14. I prefer that the teacher tell me exactly what to do and how to do it.
15. I think science is cool.
16. I think science teachers are cool.

Source: Adapted from Llewellyn 2011, p. 53.

CHAPTER 6

That said, it's important to point out that differentiation is not a synonym for "individualized instruction," the buzzword of the late 1980s (and stretching back as far as the 1960s), which demanded teachers identify and strive to attend to the unique learning needs of individual students. Given the vast cultural, behavioral, cognitive, developmental, and performance differences that characterize a "typical" class today, individualization is an overwhelming charge for a teacher.

Choice and independence emerge within and according to the parameters of class dynamics and lesson design; good science differentiation would target clusters of students with similar learning profiles. "Feasibility suggests that classroom teachers can work to the benefit of many more students by implementing *patterns* of instruction likely to serve multiple needs" (Tomlinson and McTighe 2006, p. 19; author's emphasis).

In *Differentiated Science Inquiry* (2011), Llewellyn promotes differentiation through aligning student needs and learning styles with choices for differing amounts of support and structure, or "scaffolding," provided by the teacher during activities:

> Using a differentiated inquiry approach, the teacher constructs a science investigation with multiple or tiered levels of guidance and structure so that each learner has an opportunity to choose a level that is developmentally appropriate for his or her particular learning style. Although the lesson offers various process-oriented pathways, in the end, all students arrive at the common understanding of the concept and standard being studied. (p. 29)

This last point is worth unpacking, since differentiation is sometimes reduced to adjusting outcomes based on abilities. In fact, "differentiation is seldom about different outcomes for different kids. It's about different ways to get kids where they need to go ... Or, differentiation should always be about lifting up—never about watering down" (Tomlinson, Page, and Imbeau 2013).

For some teachers, responsive or differentiated instruction requires a difficult shift away from their orientation toward "covering content," and from placing the onus for learning on the students. Responsive teachers accept responsibility for student learning, and use a variety of regular, formative assessments to monitor their progress. Assessment becomes as much a measure of teacher effectiveness in meeting the students' needs as a gauge of student content mastery.

That's not to say that traditional methodologies like summative testing and lecture are taboo, however: "There are times when explicit or direct instruction is a more appropriate choice and will complement inquiry-based teaching, especially when students have already had a great deal of direct experience with a particular phenomenon" (NRC 2007, p. 115). As Llewellyn asserts, "By having many instructional strategies available in their toolkit, teachers are most certainly in a better position to differentiate their instruction to fit the needs of their students" (2013 p. 3).

These critical attitudes and skills characterize responsive teachers:

- They accept responsibility for learner success.
- They establish clarity about curricular essentials.
- They develop communities of respect.
- They build awareness of what works for each student.
- They develop classroom management routines that contribute to success.
- They help students become effective partners in their own success.
- They develop flexible classroom teaching routines.
- They expand a repertoire of instructional strategies.
- They reflect on individual progress with an eye toward curricular goals and personal growth. (Tomlinson and McTighe 2006, p. 40)

In this list, you should see a lot of crossover with the belief and skill set of an effective teacher of good science in middle school. If a teacher is consistently strong in these attitudes and skills, Tomlinson and McTighe argue, our students will benefit, and any substantial weaknesses will correlate to learning deficits for at least some of the children who are counting on us.

Transforming Traditional Lessons

In a traditional science activity, say a pendulum lesson, the teacher and the activity's written directions define how students proceed in a step-by-step manner. How is this different from the hands-on, minds-on learning that we're advocating as good science for middle schoolers? How can you adapt a "cookbook" lesson to include science and engineering practices and inquiry strategies? Without additional materials, how can you make the necessary subtle shifts to good science?

Consider the pendulum lesson in Table 6.2 (p. 79), which shows the differences between traditional and good science instruction. This example is meant to illustrate how teachers can modify what they're already doing, rather than reinventing entire units of instruction, in order to incorporate more science and engineering practices into their daily lessons.

Other steps teachers might take toward a good science classroom include the following:

- Provide blank paper instead of worksheets and have students generate their own questions from a chapter reading assignment or teacher demonstration.
- Give students an opportunity to predict, observe, and make sense of data— essentially learn from one another as real scientists do.

- Give students a variety of materials from which to choose and explore and then encourage students to present questions about and share findings from their varied investigations before the whole class. Meanwhile, audience members can challenge the findings and probe for evidence.

Good Science Assessment

How do teachers assess learning that takes place in a science activity? In this section, we'll share the foundational ideas behind assessment of good science in the middle grades and then explore the various forms that assessment can take.

As stated in the *Framework*, the vast majority of assessment that takes place in America's classrooms today takes the shape of traditional paper-and-pencil response formats or multiple-choice tasks that are inadequate to meet the needs of good science assessment:

> Assessments of this type can measure some kinds of conceptual knowledge, and they also can provide a snapshot of some science practices. But they do not adequately measure other kinds of achievements, such as the formulation of scientific explanations or communication of scientific understanding. They also cannot assess students' ability to design and execute all of the steps involved in carrying out a scientific investigation or engaging in scientific argumentation. (NRC 2012, p. 262)

We like this concise summary of assessment that *does* align with good science methods provided by the Exploratorium Institute for Inquiry (2006):

> Assessment is part of every teacher's job. The type of assessment teachers are most familiar with—in which they examine students' work in order to determine grades, write evaluations, compare levels of achievement, and make decisions about promotion—is called summative assessment. In doing formative assessment, teachers also examine and evaluate students' thinking—but in this case, they do so in order to make pedagogical decisions for the purpose of helping students get closer to learning goals. Teachers use the information they gather about student work to determine what students need to do next that will help them progress toward the goals of the lesson. (p. 12)

Following from this general description, four premises shape our views on assessing science and engineering practices:

1. Due to the wide variations in learning styles, cognitive levels, and maturity among middle school students, we strongly recommend that teachers use pre-assessment to identify the range and nature of their students' instructional

Table 6.2. A Pendulum Lesson: Moving From a Traditional Lesson to Active Science

Traditional Approach: Understanding Pendulums	Activity-Based Approach: Understanding Pendulums
Curriculum • Presented part to whole; emphasis on basic skills • Fixed curriculum • Relies heavily on textbooks, worksheets, workbooks • *Provide materials to read and worksheet to complete regarding pendulums and motion.*	**Curriculum** • Presented whole to part; emphasis on big concepts and thinking skills • Responsive to student questions and interest • Relies on collecting primary data and using manipulative materials; students design or construct a product (pendulum) to solve the problems set forth in the focus question and assessment • *Provide each group with possible materials for making a pendulum. Ask students a focus question: "How many times does a pendulum swing?"*
Roles of Students • "Blank slates" onto which information is "etched" by the teacher • Work alone • *Students follow teacher and worksheet directions exactly: "Cut a piece of string exactly 28 cm long, and tie one end to your pencil and the other end to a washer . ..."*	**Roles of Students** • Thinkers with emerging theories about the world • Work in groups • *Ask students to construct a pendulum of their own design.*
Role of Teacher • Generally behaves in a didactic manner; disseminates information to students • Seeks the correct answer to validate student learning • *Before beginning, instruct students how to hold the pendulum and exactly how to count the swings while the teacher times them.*	**Role of Teacher** • Generally behaves in an interactive manner; guides but does not direct or mandate during the investigation; allows students to explore, make mistakes, discover on their own; provides direction as needed • Sees the students' point of view in order to understand students' present conceptions for use in subsequent lessons • *Students swing the pendulum while the teacher times them. Students arrive at different answers for how many "swings" the pendulum made in the time period. Who counted correctly? Why did that happen? Through discussion and argumentation, students discover "variables" and the need to control them.*
Assessment Is Summative • Viewed as separate from teaching; occurs almost entirely through testing • *Administer an objective, written test to determine if students understand that the length of string changes the number of swings, while the size of the bob does not.*	**Assessment Is Formative and Performance-Based** • Interwoven with teaching; occurs through teacher observations of students at work and through student exhibitions and portfolios that demonstrate what they know based on what they can do. • *Ask students to construct and demonstrate a pendulum that will swing exactly 15 times.*

Source: Adapted and updated from Cantrell and Barron. 1994.

needs and readiness to "do" good science at the start of the term and before each unit. At the start of the school year, teachers should determine the following, at a minimum: What do students know about science safety? Do they conceive of science as a process or a collection of discrete knowledge claims? Have they ever engaged in active science? (Note that we don't include a pre-assessment section in our activities, mostly due to space limitations, but we presume readers will pre-assess before engaging in activities.)

2. We are enthusiastic advocates of formative assessment. "Measuring" good science requires more than a test at the end of the unit or term to find out what students know. It should be used to inform the teacher about what is working and what isn't—*in order to continuously improve teaching and learning*. Therefore, assessment should be diagnostic, varied, and ongoing throughout the school year—formative assessment and good science go hand in hand.

3. To help students improve their thinking skills, teachers can often shift the responsibility of evaluation and taking action to the students themselves. We want students to learn how to evaluate the quality of their own findings and conclusions and make adjustments as needed based on argumentation and self-evaluation. When students are afforded this opportunity, they are thinking like scientists and involved in the highest levels of cognitive activity.

4. Pre-assessment and formative assessment constitute more than "events." They compose an instructional philosophy in which teachers assume accountability for student learning, and they are integral to authentic assessment. Teachers who constantly evaluate their students have the information necessary to modify instruction and make curricular decisions that support their students individually. In this way, testing does more than evaluate learning; it also diagnoses the effectiveness of teaching, differentiates learning, and suggests ways to improve how topics are taught.

Each of these key premises is rooted in the traits of the middle school learner, discussed in Chapter 1. Transitioning from a traditional teaching, learning, and assessing paradigm to our model entails a fair amount of self-reflection. This entails discomfort for some teachers and a genuine leap of faith for others, especially the notion that assessment says as much about the quality of instruction as the students' efforts and abilities to learn.

Formative Assessment: The Diagnosis–Prescription Cycle

"Assessment is only formative when teachers use the information they've gathered to make instructional decisions" (Exploratorium Institute for Inquiry 2006, p. 8). Good science offers continual opportunities for teachers to observe, ask questions, check skills, review procedures, and otherwise gather extensive feedback about student learning that can—and should—inform instruction and enable adjustments, clarifications, refocusing, and reteaching if necessary.

Formative assessment also strengthens differentiation. "Because formative assessment involves periodically checking students' understanding during—rather than after—instruction, it provides useful information which allows teachers to tailor their teaching to a single student's, or a whole class's, specific needs" (Exploratorium Institute for Inquiry 2006, p. 11).

Good science—and good teaching—calls for a combination of formative assessment, used to pre-assess and monitor student progress as well as the effectiveness of a teacher's instruction, and summative assessment that has dominated teacher practice for centuries. Formative measurement can be viewed as assessment *for* learning, while summative measures are assessment *of* learning; if employed properly by the teacher, formative and summative tools together are assessment *as* learning (Tomlinson, Page, and Imbeau 2013).

Along with the Preferences Survey (Figure 6.1), a diagnostic pretest at the start of the year—ungraded and nonthreatening and presented as a way to see what students know so you can avoid redundant lessons—is a way for teachers to find out what students have retained from previous science instruction. The pretest can include one or more performance-based activities; this will give the teacher a broad view of the remediation, reinforcements, and new teaching that will need to take place in the units ahead. In our pretests, we've used a variety of formats, according to the makeup of each class. Among them are the following:

- Pencil-and-paper test that surveys key concepts and terms students will need in order to master class objectives. (A version of the final unit or term test might be administered as a pretest.)

- Short essay assignment—for example, a "science autobiography" about what they know and want to know about science.

- An assignment that asks for nonverbal representations of science topics and themes, particularly for use with students with language limitations.

- Small-group activity in which students discuss science "facts" provided by the teacher—some factual and some not—and debate the merit of each "fact."

- Round-robin activity with chart paper on the wall listing a different objective on each sheet; lab teams travel from sheet to sheet to post what they know about that objective.

Ideally, a pretest series will allow a teacher to observe and evaluate the cognitive, affective, and psychomotor abilities of his or her students. A good pretest is a starting point for unit and lesson planning and should probe what students know about a variety of topics, such as safety, problem solving, key vocabulary, and equipment.

In exploring options for pre-assessment, we recommend readers investigate Page Keeley's excellent work with formative assessment probes (Keeley, Eberle, and Farrin 2005 is just one of many titles). Probes are used to examine student thinking, helping teachers understand what their students think about science and why—including common misconceptions of the sort we list in our activities—such as the notion that plants get their food from the soil (Keeley, Eberle, and Farrin 2005 p. 7). Formative assessment probes can significantly inform lesson planning and instruction. Teachers can design their own probes, or consider the examples in the extensive series of NSTA's books featuring probes designed by Keeley and her co-authors in the NSTA Learning Center.

As the school year unfolds, teachers can check student progress through multiple, ongoing means in addition to formative assessment probes—for example, journals, portfolios, lab reports, essay and research assignments, student self-assessment, observations, and performance tasks—and adjust instruction accordingly:

> One teacher, in planning a lesson on simple circuits, decided to have the students draw on the whiteboard all the circuits they tried to construct, both those that did and those that didn't work. This form of communication gave her an immediate picture of the way the students' ideas were developing and enabled her to work with those who were unsure and needed help understanding what is essential in a complete circuit. (Harlen 2003, p. 22)

Ongoing formative assessment used to diagnose shortcomings in student learning is a window on the effectiveness of a teacher's lessons as well and should be used to identify improvements in teaching. Using a variety of assessment types, like using a variety of instructional methods, provides a fuller profile of student learning and is a key to good science in the middle school.

The *NGSS* are themselves assessments—performance expectations—calling on students to demonstrate what they know through means other than the multiple-guess format that has come to dominate K–12 assessment in other subjects. Consider Table 6.3, which lists a sampling of ideas teachers can use to evaluate scientific understanding.

Table 6.3. Assessing Student Learning in Science and Engineering: Examples

Constructed Response	Performance-Based Assessment
• Science lab notebook • Diagram, graph, or table • Podcast or digital media project • Concept map • Science-themed story or play • Science song • Graphic novel • Artwork • Collage or scrapbook • Research report	• Use models to demonstrate or explain • Conduct specific science tasks (e.g., use balance or probeware correctly) • Design an investigation to answer a question or solve a problem • Lab proficiency • Musical, artistic, or dramatic performance • Conduct an experiment • Debate • Computer-based simulation • Reciprocal teaching • Oral presentation

The diagnosis-prescription cycle is unending for effective middle-grade teachers. The goal is to appraise the range of skills and aptitudes that are found among middle schoolers, while monitoring both student learning and instructional effectiveness. We reiterate that effective assessment "is about before, during, and after—not just after" (Tomlinson, Page, and Imbeau 2013).

In *Meet Me in the Middle* (2001), Wormeli writes about his experience with the kinds of assessment methods that we have described above. While some assignments might appear "soft" or too simplistic, he says, effective manipulation of content in these methods requires considerable skill on the part of middle-grade students. Wormeli describes his success giving a class presentation about how cells die by conducting a funeral for a dead cell, which he deftly entitled "Death of a Cellsman." Wormeli offers additional alternatives to traditional assessments, all of which can be used at the middle school level in any subject:

- Journal or diary entries
- Radio plays
- Video productions
- Annotated catalogs of artifacts
- Games and puzzles
- Museum guides
- Historical or science fiction stories
- Almanacs

- Interview with an expert
- News or feature articles
- Timelines or murals
- Advertisements. (2001, p. 97)

Metacognition as an Assessment Outcome

The important goal of fostering metacognition—helping students develop an awareness of the quality of their own thinking and course work—is an assessment outcome teachers can achieve in the middle grades by employing some guided questioning. Here are three important prompts from our first edition that teachers tell us they still find useful for fostering metacognition with middle schoolers:

1. Where are you trying to go? (Identify and communicate the learning and performance goals.)
2. Where are you now? (Assess, or help the student self-assess, current levels of understanding.)
3. How can you get there? (Help the student with strategies and skills to reach the goal.) (Atkin, Black, and Coffey 2001, p. 14)

Students can revisit these three questions, or variations of them, by keeping a learning log to chart their progress and reflect on their own discoveries.

For many teachers, assessing active learning using formative methods is a departure from their preparation and practice, but it's a shift that's embodied in the *NGSS* construct for assessment and a powerful complement to doing good science.

References

Atkin, J. M., P. Black, and J. Coffey, eds. 2001. *Classroom assessment and the national science education standards.* Washington, DC: National Academies Press.

Cantrell, D., and P. Barron, eds. 1994. *Integrating environmental education and science.* Newark, OH: Environmental Education Council of Ohio.

Exploratorium Institute for Inquiry. 2006. Introduction to formative assessment facilitator's guide. *www.exploratorium.edu/ifi/workshops/assessing*

Harlen, W. 2003. *Enhancing inquiry through formative assessment.* San Francisco: Exploratorium.

Keeley, P., F. Eberle, and L. Farrin. 2005. Formative assessment probes: Uncovering student ideas in science. *Science Scope* 28 (5): 18–21.

Llewellyn, D. 2011. *Differentiated science inquiry.* Thousand Oaks, CA: Corwin Press.

Llewellyn, D. 2013. Choice: The dragon slayer of student complacency. *Science Scope* 36 (7): 90–95.

National Research Council (NRC). 2007. *Taking science to school: Learning and teaching science in grades K–8.* Washington, DC: National Academies Press.

National Research Council (NRC). 2012. *A framework for K–12 science education: Practices, crosscutting concepts, and core ideas.* Washington, DC: National Academies Press.

Tomlinson, C. A. 2008. The goals of differentiation. *Educational Leadership* 66 (3): 26–30.

Tominson, C. A., and J. McTighe. 2006. *Integrating differentiation and understanding by design: Connecting content and kids.* Alexandria, VA: Association for Supervision and Curriculum Development.

Tomlinson, C. A., S. Page, and M. Imbeau. 2013. Differentiation and 21st-century skills. ASCD Professional Development Conferences and Institutes: March 13–15. Chicago, IL.

Wormeli, R. 2001. *Meet me in the middle: Becoming an accomplished middle-level teacher.* Portland, ME: Stenhouse.

Overview of the 10 STEM Activities for Middle School Science

Differentiated, Standards-Based, and Fun!

Before we move to the activities, we need to give a disclaimer or two about using any template for lesson planning. The components of the lesson-plan template presented in these activities are all integral to good science instruction, but, as teachers know, units and lessons need to be tailored to the specific needs of different groups of students, different days, and even variations in unit objectives. Teachers using our template are encouraged to modify it as needed to fit their own circumstances. (See more about making modifications in Chapter 6.)

We know that the activities are lengthy and might look cumbersome, but we've opted to provide extensive detail so that they are truly ready-to-use.

Let us reiterate that good science lessons should be designed with practicality and instructional effectiveness in mind. It is frankly not feasible to do hands-on, minds-on activities all the time, in part because the setup, process, and cleanup are time intensive. Nor is it prudent from a pedagogical standpoint. Science instruction should be varied—structured to incorporate a range of strategies that will engage different learning styles.

Also, readers should be aware that although the activities in this chapter are offered as examples of good science, they are in no way intended to represent a complete middle school science curriculum for any particular state or school district. Our activities emphasize the three dimensions of the *Framework*: scientific and engineering practices, crosscutting concepts, and disciplinary core ideas, with English language arts and STEM integrations. Our 10 activities provide a context for further study of biology, geology, chemistry, physics, engineering, and astronomy. They are examples of what's possible; we hope you find them to be useful and vivid illustrations of what good science in middle school can be.

Content- and Kit-Based Instruction

The *NGSS* content standards provide a framework for teaching the disciplinary core ideas—life sciences, Earth and space sciences, and engineering and technology—as well as the crosscutting concepts:

1. Patterns
2. Cause and effect: Mechanism and explanation
3. Scale, proportion, and quantity

4. Systems and system models
5. Structure and function
6. Energy and matter: Flows, cycles, and conservation
7. Stability and change (NGSS Lead States 2013)

In those states that have adopted the *NGSS*, state standards and district curricula should reflect the *NGSS* content standards and crosscutting concepts with a corresponding emphasis on science and engineering practices, including:

1. Asking questions (for science) and defining problems (for engineering)
2. Developing and using models
3. Planning and carrying out investigations
4. Analyzing and interpreting data
5. Using mathematics, information and computer technology, and computational thinking
6. Constructing explanations (for science) and designing solutions (for engineering)
7. Engaging in argument from evidence
8. Obtaining, evaluating, and communicating information

Many school systems around the country that have implemented good science use kits in their programming because they remove a number of roadblocks to implementing active science and applying science and engineering practices.

We define *kits* broadly to include a range of products, from district- and school-developed materials to commercially produced curricula. Kits contain materials necessary for students to conduct scientific investigations and experiments organized into units spanning several weeks. The units usually consist of 10 or more lessons and are structured according to teacher guides. While many educators and some school systems develop and refurbish their own kits, units for middle school instruction may also be purchased from a number of vendors.

Kits began as a not-for-profit venture on the part of teachers who wanted to share an excellent lesson and boxed the self-contained student activities. Subsequently, the demand for kits led to their creation and distribution by curriculum marketers.

Commercial kits claim to be aligned with standards, and our experience mostly supports that claim; however, it is the teacher's responsibility to track the kit activities against the standards they aim to satisfy. For more guidance, at local workshops and large conferences kit vendors train teachers to become familiar with unit activities and to lead students through the materials with a minimum of direct instruction. (See Chapter 18 for information about vendors of kit-based instruction.)

While kits can facilitate good science, it's important for us to point out that the science isn't *in* the kit; it emerges *from* the kit. That is, beyond using the engaging activities contained in a boxed science unit, the teacher is responsible for framing the problem and setting up focus questions, guiding the practices as a co-investigator, helping students make connections to important concepts and key terminology, and constantly monitoring learning to assess the need for reinforcement, review, and enrichment.

Another Safety Reminder

While we devoted most of Chapter 5 to science classroom safety, we must emphasize here that teachers engaging in activity-based science must be properly trained and well prepared for developing and conducting safe investigations, in the context of a safe learning environment. Teachers who lack confidence in science safety procedures and standards are urged to study Chapter 5, for starters, and to explore the safety resources referenced in Chapter 18, in addition to workshops and trainings provided by their districts or state science teacher associations.

Our 5E Activity Template

There are variations of this book's 10 sample activities in use across the country, and we certainly don't attempt to illustrate all of them in this chapter. We aligned our activities with the *NGSS* and *Common Core State Standards*, so they're ready for you to use now. The template itself is adaptable to a variety of STEM lessons—challenging students, according to the 5E instructional model—to engage, explore, explain, elaborate, and evaluate (Bybee 2006; see Chapter 3 for a discussion of the 5E instructional model). Below we describe each section of our activity template.

Safety First

This section of the activity template is a standing reminder of the primary importance safety must play in all lesson and activity planning. We can't say it often enough, so we'll say it first in every activity.

STEM

This section identifies the STEM dimensions—science, technology, engineering, and math—engaged in the investigation. As we discuss in Chapter 4, while a "definition" of

STEM is elusive, for the purposes of these activities, a STEM lesson is one that includes at least two of the four disciplines in an integrated manner. Engineering design is prominent in the *NGSS*, and while it may be a new topic for many teachers, we find it helpful to view engineering design as the application of scientific knowledge to solve problems. In this context, engineering provides a natural next step for students to more deeply understand science as the means to solving problems facing human beings. "Engineering as problem solving" is integral to STEM, rather than a separate "kind" of science.

Tying It to the Standards

This section identifies which *Next Generation Science Standards* and *Common Core State Standards, English Language Arts* and *Mathematics* are explicitly met by the activity. We also lay out the disciplinary core ideas and crosscutting concepts from the *Framework* and *NGSS* for our activities to provide convenient references for you.

Misconceptions

This section lists common misconceptions held by students (and sometimes teachers) about the science topics and practices covered in the activity, and one or two recommended resources available to learn more.

Objectives

This section states the learning objectives for the lesson, listed in terms of what the students will do, know, and/or demonstrate as a result of the activity or investigation.

Academic Language

This section lists the vocabulary with which students will become familiar as they move through the activity.

Focus Question(s) for Scientific Inquiry Activities

Focus questions for inquiry activities can be used as a form of frontloading, as a way to engage students and generate discussion with the whole class, or as "tasks" assigned to collaborative table groups (see Chapter 2) to initiate investigations. Investigations start with a question or problem. Sometimes it's a question or problem posed by the teachers, especially with younger middle schoolers new to activity-based science; other times (later in the school year or with older students) it's a question or problem that students generate or encounter themselves.

Identifying the Design Problem(s) for Engineering Activities

Generally the teacher provides the design problem for students to solve. Teachers can differentiate according to the difficulty of the problem. You may choose to have students develop, test, and solve their own problems as well, although this may require careful guidance to ensure that a problem is properly defined and framed. Teachers designing engineering activities are urged to pay attention to safety issues, including engineering controls, safety procedures, and personal protective equipment required to minimize the risk of these sorts of activities.

Teacher Background

This section provides information teachers can use to help them when preparing for the activity, generally going beyond what students need to know

Preparation and Management

This section provides the nuts and bolts of the activity. In addition to the prep time and the teaching time, this section also lists materials associated with the activity so users can gauge the activity's costs.

5E Instructional Model

This section enables teachers to identify which of the skills (the "E"s) are activated in each activity and/or which are emphasized. See our detailed explanation of the 5E components in Chapter 3, and note that it is not necessary or expected for teachers to cover all five Es in one class period.

Discussion and Argumentation

Discussions of good science activities should focus on *argumentation*, wherein students defend and justify the claims they make from evidence, verbally and in writing, after analyzing their data, and respond to questions and challenges from their audience. This is where students engage in risk taking and demonstrate scientific reasoning skills, thinking like scientists and engineers and experiencing the *passion* of good science! For a list of prompts teachers can use to engage students in argumentation, see Figure 2.3, p. 17; also refer to Activity 8, "Saving the World—One Ecosystem at a Time" (p. 173), which emphasizes argumentation.

Differentiation

This section lays out variations on the activity and assessment ideas to reach students who need more support ("Broader Access/Modified") or more challenge ("Extension Activity"/"Challenge Assessment").

Note: Our titles for these options are for teacher reference only; teachers are urged to rename the activity and assessment options to avoid any impression of labeling students.

Sample Activities

Variations of the following 10 activities have been successfully used by teachers all over the country, working with a wide range of students of different ages, demographics, and ability levels. We've modified and adapted them to suit the needs of our students, just as you will. (There are many resources listed in Chapter 18 that will lead you to more lesson ideas.)

Please note that our activities are intentionally varied and all reflect recommended instructional practices. As you test these activities for yourself and try these activities with your students, feel free to mix and match strategies across the activities to provide the most effective match for you and your students. We have attempted to provide sufficient support for teachers new to teaching science, while also including opportunities for all readers to take these activities as starting points and adapt to their own strengths and interests in reaching the students in their classes.

How do we transition to the activities themselves? You guessed it! Before moving to the activities, let us reiterate how important it is that teachers provide adequate safety instruction, procedures, and persistent safety reminders for students to follow, and closely monitor students who are conducting the investigation. (See Chapter 5 for more on good science safety.)

Here is a list of the activities according to science and engineering disciplines:

Activity 1:   Physical science: Magnetism
Activity 2:   Physical science: Energy transfer
Activity 3:   Engineering: Structural design
Activity 4:   Earth science: Natural disasters and engineering technology
Activity 5:   Biology: Genetic variation
Activity 6:   Earth science: Water conservation and human impact on Earth systems
Activity 7:   Biology: Population dynamics/natural selection

Activity 8:   Biology: Ecosystems and biological diversity

Activity 9:   Science and engineering processes: Testing and communication

Activity 10:   Earth science: Weather and climate

We expect that readers will pick and choose from the collection, so we designed each activity to be used independently from the others. Consequently, in addition to the safety reminder at the start of each activity, you'll find that we intentionally repeat key recommendations or resources throughout the activities.

Enjoy your foray into the 10 activities!

Please note: Remember that these good science activities are just a few of hundreds, maybe thousands, out there. Refer to Chapter 18 for resources and leads to more activities.

References

Bybee, R., et al. 2006. The BSCS 5E instructional model: Origins and effectiveness: A report prepared for the office of science education, National Institutes of Health. Colorado Springs, CO: BSCS.

NGSS Lead States. 2013. *Next Generation Science Standards: For states, by states*. Washington, DC: National Academies Press. *www.nextgenscience.org/ next-generation-science-standards*.

CHAPTER 8

Activity I: Invisible Force Fields
Physical Science: Magnetism

Have you ever heard someone say, "I'll believe it when I see it"?

Can you see everything that is important in the STEM disciplines? Or are there things you can't see? If you can't see it, how do know something exists? How do you learn about it?

We don't see the wind, but we can see dust, leaves, and other visible objects that the wind moves. Understanding how we learn about things we cannot see is an important part of basic scientific investigation. This activity places students in the role of scientists and engineers as they learn about magnetic fields and use their new information to design a smaller and stronger electromagnet.

Safety First

In every activity, we remind you to be certain that you understand the potential risks involved and are confident you can ensure your students' safety. Before attempting any of these activities in class, we recommend completing them yourself and optimally with a teaching partner.

STEM

This is a great activity to engage students in the practices associated with testing hypotheses and determining characteristics and properties of phenomena that they cannot see directly. The practice of inference is a critical step toward innovation and creativity for all learners as they make claims based on evidence.

This activity encourages students to investigate and develop their scientific knowledge of magnetic fields. Then, students use their scientific knowledge to design a piece of technology (the smallest and strongest electromagnet possible with the supplies available according to design parameters).

Tying It to the *NGSS*

Ask questions about data to determine the factors that affect the strength of electric and magnetic forces. (MS-PS2-3)

[Clarification Statement: Examples of devices that use electric and magnetic forces could include electromagnets, electric motors, or generators. Examples of data could include the effect of the number of turns of wire on the strength of an electromagnet, or the effect of increasing the number or strength of magnets on the speed of an electric motor.]

[Assessment Boundary: Assessment about questions that require quantitative answers is limited to proportional reasoning and algebraic thinking.]

Conduct an investigation and evaluate the experimental design to provide evidence that fields exist between objects exerting forces on each other even though the objects are not in contact. (MS-PS2-5)

[Clarification Statement: Examples of this phenomenon could include the interactions of magnets, electrically charged strips of tape, and electrically charged pith balls. Examples of investigations could include firsthand experiences or simulations.]

[Assessment Boundary: Assessment is limited to electric and magnetic fields, and limited to qualitative evidence for the existence of fields.]

Science and Engineering Practices

Asking Questions and Defining Problems

Ask questions that can be investigated within the scope of the classroom, outdoor environment, and museums and other public facilities with available resources and, when appropriate, frame a hypothesis based on observations and scientific principles. (MS-PS2-3)

Asking questions and defining problems in grades 6–8 builds on grades K–5 experiences and progresses to specifying relationships between variables, clarifying arguments and models.

Ask questions to identify and clarify evidence of an argument. (MS-ESS3-5)

Planning and Carrying Out Investigations

Plan an investigation individually and collaboratively, and in the design: identify independent and dependent variables and controls, what tools are needed to do the gathering, how measurements will be recorded, and how many data are needed to support a claim. (MS-PS2-2)

Conduct an investigation and evaluate the experimental design to produce data to serve as the basis for evidence that can meet the goals of the investigation.

Plan for safety by determining which engineering controls, safety procedures, and personal protective equipment will be needed. (MS-PS2-5)

Constructing Explanations and Designing Solutions

Apply scientific ideas or principles to design an object, tool, process, or system. (MS-PS2-1)

Engaging in Argument From Evidence

Construct and present oral and written arguments supported by empirical evidence and scientific reasoning to support or refute an explanation or a model for a phenomenon or a solution to a problem. (MS-PS2-4)

Disciplinary Core Ideas

PS2.B: Types of Interactions

Electric and magnetic (electromagnetic) forces can be attractive or repulsive, and their sizes depend on the magnitudes of the charges, currents, or magnetic strengths involved and on the distances between the interacting objects. (MS-PS2-3)

Forces that act at a distance (electric, magnetic, and gravitational) can be explained by fields that extend through space and can be mapped by their effect on a test object (a charged object, or a ball, respectively). (MS-PS2-5)

Crosscutting Concepts

Cause and Effect

Cause and effect relationships may be used to predict phenomena in natural or designed systems. (MS-PS2-3), (MS-PS2-5)

Systems and System Models

Models can be used to represent systems and their interactions—such as inputs, processes, and outputs—and energy and matter flows within systems. (MS-PS2-1), (MS-PS2-4)

Stability and Change

Explanations of stability and change in natural or designed systems can be constructed by examining the changes over time and forces at different scales. (MS-PS2-2)

Tying It to *Common Core State Standards*, ELA

RST.6-8.3 Follow precisely a multistep procedure when carrying out experiments, taking measurements, or performing technical tasks. (MS-PS2-1), (MS-PS2-2), (MS-PS2-5)

WHST.6-8.1 Write arguments focused on discipline-specific content. (MS-PS2-4)

WHST.6-8.7 Conduct short research projects to answer a question (including a self-generated question), drawing on several sources and generating additional related, focused questions that allow for multiple avenues of exploration. (MS-PS2-1), (MS-PS2-2), (MS-PS2-5)

Tying It to *Common Core State Standards, Mathematics*

MP.2 Reason abstractly and quantitatively. (MS-PS2-1), (MS-PS2-2), (MS-PS2-3)

7.EE.B.4 Use variables to represent quantities in a real-world or mathematical problem, and construct simple equations and inequalities to solve problems by reasoning about the quantities. (MS-PS2-1), (MS-PS2-2)

Common Misconceptions

There are many misconceptions about electricity and magnetism. One common misconception is that electricity is a form of energy; another is that electricity is the flow of electrons. Both of these statements are false, and illustrate the importance of using words correctly as we discuss and help students learn disciplinary core ideas and crosscutting concepts.

These misconceptions emerge from various sources, including textbooks and other teacher resources that provide only a surface explanation without paying attention to how the authors use technical vocabulary.

If you search the internet, you will find many websites that provide a range of explanations and accurate representations regarding magnetism and electricity. We would need many pages to provide the requisite factual scientific information regarding electricity, electrical energy, electric current, and magnetism.

Therefore, we suggest using the free Science Object from the NSTA Learning Center entitled "Electric and Magnetic Forces: Electromagnetism" (*http://learningcenter.nsta. org*). Select "middle school," and then "free resources."

Dr. Philip Sadler and others associated with the Harvard-Smithsonian Center for Astrophysics have produced a series of videos and resources for science and mathematics educators regarding how to work with students to help them break through their misconceptions. You can find "Minds of Our Own" and other free online resources and video resources at *www.learner.org/resources/series26.html*.

Another excellent resource to help determine what students know (and don't know) is Page Keeley's series *Uncovering Student Ideas in Science*: *http://uncoveringstudentideas.org*.

Objectives

By the end of this activity, students will have demonstrated the ability to:

- Test hypotheses regarding magnetic fields.
- Show their scientific knowledge regarding the presence of magnetic fields based on direct and indirect observation and experimentation.
- Design, build, and test an electromagnet within design parameters.
- Engage in scientific argumentation based on empirical data.

Academic Language

Magnet, magnetic field, compass, electromagnet, empirical, scientific knowledge, inference, parameter, argumentation, data

Focus Question(s) (Scientific Inquiry)

- What are the effects caused by a magnet?
- How can we test for the presence of a magnet or magnetic field?
- What can be affected by a magnet? What's *not* affected?
- What determines how powerful a magnet is?

Framing the Design Problem(s) (Engineering Practice)

Using the materials provided by your teacher, design, build, and test the lowest mass and strongest electromagnet. The successful design will minimize the ratio of the mass of the electromagnet compared to the maximum mass lifted to a height of 15 cm by the electromagnet.

Refer to Figure 4.3 "The Engineering Design Process" (p. 45) for more information about how to frame the design problem for this activity.

Teacher Background

To review content knowledge associated with electromagnets and magnetic fields, we suggest teachers research the sources listed under the "Common Misconceptions" section; there also are many good YouTube and Khan Academy videos and TEDed resources to consider as well. Here are some specific pieces that some of you may find helpful as background information:

- The ratio of mass to strength is expressed mathematically as electromagnet mass/mass lifted = ratio.

- *Strength* is defined in this situation as the mass of washers lifted to a height of 15 cm, in typical classroom atmosphere air and in normal classrom conditions.
- For example: An electromagnet has mass of 25 grams and lifts 9 small washers with a mass of 0.15 grams. The ratio would be 25 g /0.15 g = 166.67. For a second electromagnet with a mass of 5 grams that lifts 9 small washers with a mass of 0.15 grams, the ratio would be 5 g/0.15 g = 33.33. Since the lower mass electromagnet lifted the same mass of washers, the lower mass electromagnet is determined to be more "efficient" or more desirable in this contest. In this example, the ratio demonstrates the mathematical relationship of ratios and fractions and when we want to minimize a ratio to "win" a design challenge.

Optimize this activity by allowing students to develop explanations following from their experiences with the electromagnets and share their claims based on the evidence they collect. Then, the teacher can help students reach a common understanding by considering the evidence collected as a class data set.

Once the students have demonstrated a basic understanding of electromagnets and magnetic fields based on their investigation (as demonstrated through their evidence-based argumentation), the next step is presenting a design challenge to the students. The design challenge phase of this activity should be left as open as possible to increase the opportunity for student creativity and innovation in developing a solution to the challenge.

Students should be allowed to develop their own data collection plan and determine how to analyze the data sets. Let the students know that they'll be defending their plan using whiteboards in a sharing setting.

Preparation and Management

Gather the materials for the student work group kits. Each student workgroup will need the following:

Item	Source and/or Specifications
30 small washers	SAE zinc #6 Home Depot Item# 030699197910
1 nail (2d size is OK)	You could also use a zinc machine screw, #6-32 2", Home Depot Item# 030699275816
1 two "AA" battery holder (with leads)	Radio Shack Item: 2700408
1 40 cm length of insulated wire; 15mm striped in each end	22-gauge stranded hookup wire, Radio Shack #2781224
1 60 cm length of insulated wire; 15mm striped in each end	22-gauge stranded hookup wire, Radio Shack #2781224

2 AA batteries	Do not use rechargeable batteries. They may get very hot, and could burn a student.
1 small paper clip with 15 cm thread tied to one end	You will want to use the thinnest thread you can find.
2 small (3/4 in. diameter) disk ceramic magnets	Home Depot Item #03069997043 package of 8
Safety glasses or goggles	Commercial science supplier (PPE must meet ANSI Z87.1 standard)

(All items, except the safety goggles, will fit inside an 8 oz. margarine tub with a snap-on lid to be used again in the future.)

Prep Time

1–3 hours to gather materials (depending on availability of materials)

60 minutes to construct the kits student teams will use for the Engage, Explore, and Design Challenge stages of the activity

Teaching Time

Two to four teaching periods (generally 45 minutes per teaching period)

Materials

- Magnets
- Small nails (1 per lab set; 2d common nails available at most hardware or home improvement stores)
- Battery clips (1 per lab set, size AA; you will want to have both single battery and double battery clips available)
- AA batteries (2 per lab set; do not use rechargeable batteries, as they generate too much heat)
- 22-gauge insulated copper wire (65 cm length, striped 15 mm on each end)
- #6 zinc washers (30 per lab set)
- Small paper clip (1 per lab set)
- Thread (65 cm per lab set, sewing thread, the thinner the better, strong enough to hold a small paper clip)
- Safety glasses or goggles
- A balance will be needed during the Elaborate and Design Challenge stages of the lesson.

5E Instructional Model

Engage

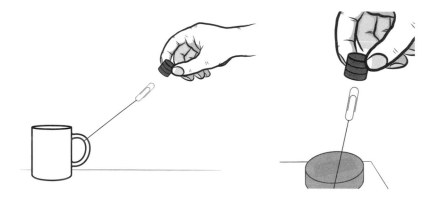

Put on safety glasses or goggles. Demonstrate the floating paper clip (see drawings). Using a large paper clip, tie a thin thread to the paper clip at one end, and to an "anchor" that will rest on a tabletop. Use multiple magnets to generate a sufficient magnetic field to "float" the paper clip at the end of a thin thread. With a few minutes of practice, you should be able to demonstrate that the magnet field will attract the paper clip and allow you to have "open space" between the magnet(s) and the paper clip.

When working with students, you could have them hold a piece of paper between the magnets and the paper clip to demonstrate that the paper does not have a perceptible effect on the "invisible force" that must be acting on the magnet. When you remove the magnet, the paper clip falls back to the table top.

Questions for the Engage stage:
- What do you see?
- Does the clip stay 'up' when the magnets are moved away?
- How far away can you move the magnets, and continue to have the paper clip "float" in the air?
- What happens when you put one thickness of paper between the magnets and the paper clip?

Preassessment Questions
- How can you explain this phenomenon?
- What evidence do you have for your explanation?

These questions may be given as writing prompts and reviewed to see how and what students think about magnetism as well as their fluency with argumentation. By gauging their prior knowledge and skills, you'll be better able to shape subsequent instruction around their needs and ability levels.

Explore

Tell the students that they will be given a "kit" to explore the phenomenon more thoroughly. Give the students the kit and have them experiment with the items in the kit to explore everything they can about the phenomenon.

Explain

Each team of scientists should produce a whiteboard set of "claims and evidence" regarding the phenomenon of the floating paper clip. These should be shared with the class. The teacher facilitates the discussion, including many opportunities for students to question and review claims and evidence. The teacher and students may ask standard questions such as "How do you know?" and "What did you see that would support your statements in your explanation?"

Following the discussion, you may suggest students look on the internet or use other classroom resources to locate additional information on the topic that supports, refutes, clarifies, or obscures the students' understanding of what could be happening to explain the observations. (All of this constitutes formative assessment data for the teacher to use to target instruction and further investigation into the phenomenon.)

After each group has defended its information, findings should be written up for comparison to a whole-class set of common claims and evidence. The teacher elicits (or provides, depending on the class and available time) additional questions that would need to be answered for any variation between the class set of common claims and the claims each team made based on their evidence.

Note that when whiteboarding like this, it is important to first establish norms with the class. For example, say, "Every group member must contribute to the explanation or answer questions from other classmates," or "One person does not carry the weight for everyone. No slackers!"

Elaborate (Engineering Design Challenge)

Students are set on the task of designing an electromagnet that minimizes the mass ratio of the electromagnet to the mass listed by the electromagnet. The students will need to provide a set of data to support their claims. Additionally, students will need to provide a written elaboration of their mass ratio of the electromagnet to the mass lifted by the electromagnet. Included in their written elaboration should be the decisions they made (and the rationale for the decisions) that led them to arrive at their design solution. Elaborations must include mathematical representations as well as narrative and labeled diagrams. You will need to provide a balance to measure the mass of the electromagnets.

Evaluate

Students are evaluated based on the depth of understanding expressed in their Exploration as well as Elaboration phases. The team with the smallest mass to strength ratio doesn't necessarily earn the best marks in the evaluation. The evaluation should also be based on the quality of detail and alignment between claims and evidence, and the elaboration information from the Engineering Design Challenge. Make sure that students know the evaluation criteria in advance.

Discussion and Argumentation

Students should be given multiple opportunities to voice their claims and evidence. They should also be given the opportunity to refine their claims based on class-pooled evidence. This dialogue is facilitated by the teacher, and should focus on the nature of scientific ways of knowing (systematic review of data and the processes used to collect the data).

Scientists revise their findings all the time when they gain new information! We are not in search for "one correct answer." Instead, we suggest the students stay focused exploring and articulating what they "know" and the evidence they have. If the students need to go back to the activity to run additional tests, that should be encouraged. The audience should be encouraged to challenge the team presenting with pertinent questions, or to challenge their findings if they come up with different data. Of course, all of this depends on how much time can be allocated to this activity. (Note: There is a growing awareness for the need to manage a "safe" learning environment for every learner in the classroom community. Over time, many teachers are able to support the development of a classroom learning community where challenges are offered to strengthen the community understanding, and everyone in the classroom is safe to "risk" making mistakes and everyone is encouraged to offer suggestions and clarifying questions to refine, affirm, and suggest revisions to claims based on evidence.)

If Time Allows

One of the many important opportunities provided in this activity is the chance to help students experience the analysis and redesign aspects referenced in the *NGSS* domain of science and engineering practices. This can be accomplished in less time than the first cycle of this lesson. Students already have a familiarity with the materials and processes. Taking an additional 45 minutes can lead to amazing advances in student learning and understanding.

We suggest that you review the findings of the class and allow 1–2 days to pass to allow students to use out-of-school time to explore (safe!) options regarding improving their design and manufacture of a new improved electromagnet. This will not cost any additional instructional time, except reminders about how to do any out-of-class experimenting safely, and will allow students to create and innovate options in electromagnetism. Then, you can have a 45-minute session for students to construct and analyze data from another cycle of investigation starting with Explore, and then moving into Explain, Elaborate, and Evaluate.

Differentiation

1. **Broader Access Activity:** Limit the items provided in the kit at each stage of the 5E lesson plan. By limiting the diversity of the items, students who are easily distracted would have their attention focused only on the relevant items necessary for a particular stage of the activity.

2. **Extension Activity:** Expand the options and variables among the items provided in the kit at each stage of the 5E lesson plan (various lengths and diameter of wire, additional battery sets, larger nails, smaller nails, larger washers, stainless steel items, etc.). By expanding the choice of items, students who want to try additional options could test and design to the same specifications and have a greater opportunity to elaborate and explore.

3. **Modified Assessment:** Modify the assessment to focus on the verbal, written, or diagrammatic elaboration of the principles in this lesson.

4. **Challenge Assessment:** Allow students to present a comparative analysis of the effects they measured among the supplies mentioned in the Extension Activity. This could be an oral presentation, diagram or artwork, written report, or other format.

Note: These are examples as illustrations of possible differentiation options. The actual adaptations you create will depend on the results of your pre-assessment and ongoing formative assessments of individual students.

Teacher Content Note: We suggest teachers clarify the use the word "mass" as a noun and a verb. In this activity, we used the word "mass" as a verb and noun. We have heard many teachers use the noun "weight" or verb "weigh" when working with objects in the lab. This can lead to misconceptions related to forces and motion, and Newton's Laws of motion.

We acknowledge that under usual conditions from our individual frame of reference here on earth, weight and mass are easily confused. This often happens when we use everyday language and tools in the science lesson. For example, a bathroom scale often uses a spring or force measuring device to measure the force of our body upon the surface of the bathroom scale. And this reads out in pounds, which is a unit of force. Yet, we often convert pounds into kilograms or grams, which are a unit of mass. And mass is measured using a balance. If we take the balance and objects into space or onto another planet, the balance will continue to give the same reading for an object regardless of whether you are on Earth or another plant traveling at a consistent velocity. However, a bathroom scale will read a different weight for the same object. This inconsistency can be demonstrated in an elevator here on Earth. If you stand on a balance (like the ones used in many doctors' offices) in an elevator, the balance will remain consistent in reading the mass of your body. However, if you use a bathroom scale to get your "weight" in an elevator, you will see that the reading changes as the acceleration of the elevator car changes as it starts and stops. For more information on this, you may check website resources such as TEDed lessons (*www.ted.com*), Kahn Academy (*www.khanacademy.org*), or materials available through the NSTA Learning Center (*http://learningcenter.nsta.org*).

Activity 2: Keeping Your Cool
Physical Science: Energy Transfer

What do rattlesnakes, kangaroo rats, and teenagers have in common? No, this is not a joke: All three species go to extraordinary measures to stay cool during the dog days of summer.

This activity gets students thinking about how heat is transferred through conduction, convection, and radiation. After collecting preliminary data, a Design Challenge invites students to figure out the best shade structure design to keep their cool on a hot summer day.

Safety First

In every activity, we remind you to be certain that you understand the potential risks involved and are confident you can ensure your students' safety. Before attempting any of these activities in class, we recommend completing them yourself and optimally with a teaching partner.

STEM

This activity involves learning the science of energy transfer, using technology to collect data, and enhancing math skills while designing a product. STEM is incorporated into all aspects of this activity.

Tying It to *NGSS*

Apply scientific principles to design, construct, and test a device that either minimizes or maximizes thermal energy transfer. (MS-PS3-3)

[Clarification Statement: Examples of devices could include an insulated box, a solar cooker, and a Styrofoam cup.]

[Assessment Boundary: Assessment does not include calculating the total amount of thermal energy transferred.]

Science and Engineering Practices

Planning and Carrying Out Investigations

Planning and carrying out investigations to answer questions or test solutions to problems in grades 6–8 builds on K–5 experiences and progresses to include investigations that use multiple variables and provide evidence to support explanations or design solutions.

Plan an investigation individually and collaboratively, and in the design identify independent and dependent variables and controls, what tools are needed to do the gathering, how measurements will be recorded, and how many data are needed to support a claim. (MS-PS3-4).

Constructing Explanations and Designing Solutions

Constructing explanations and designing solutions in 6–8 builds on K–5 experiences and progresses to include constructing explanations and designing solutions supported by multiple sources of evidence consistent with scientific ideas, principles, and theories.

Apply scientific ideas or principles to design, construct, and test a design of an object, tool, process, or system. (MS-PS3-3)

Plan for safety by determining which engineering controls, safety procedures and personal protective equipment will be needed.

Engaging in Argument From Evidence

Engaging in argument from evidence in grades 6–8 builds on K–5 experiences and progresses to constructing a convincing argument that supports or refutes claims for either explanations or solutions about the natural and designed worlds.

Construct, use, and present oral and written arguments supported by empirical evidence and scientific reasoning to support or refute an explanation or a model for a phenomenon. (MS-PS3-5)

Disciplinary Core Ideas

PS3.A: Definitions of Energy

Temperature is a measure of the average kinetic energy of particles of matter. The relationship between the temperature and the total energy of a system depends on the types, states, and amounts of matter present. (MS-PS3-3), (MS-PS3-4)

PS3.B:

PS3.B: Conservation of Energy and Energy Transfer

When the motion energy of an object changes, there is inevitably some other change in energy at the same time. (MS-PS3-5)

The amount of energy transfer needed to change the temperature of a matter sample by a given amount depends on the nature of the matter, the size of the sample, and the environment. (MS-PS3-4)

Energy is spontaneously transferred out of hotter regions or objects and into colder ones. (MS-PS3-3).

ETS1.A: Defining and Delimiting an Engineering Problem

The more precisely a design task's criteria and constraints can be defined, the more likely it is that the designed solution will be successful. Specification of constraints includes consideration of scientific principles and other relevant knowledge that is likely to limit possible solutions (secondary to MS-PS3-3).

ETS1.B: Developing Possible Solutions

A solution needs to be tested, and then modified on the basis of the test results in order to improve it. There are systematic processes for evaluating solutions with respect to how well they meet criteria and constraints of a problem (secondary to MS-PS3-3).

Crosscutting Concepts

Energy and Matter

Energy may take different forms (e.g., energy in fields, thermal energy, energy of motion). (MS-PS3-5)

The transfer of energy can be tracked as energy flows through a designed or natural system. (MS-PS3-3)

Tying It to *Common Core State Standards*, ELA

RST.6-8.1 Cite specific textual evidence to support analysis of science and technical texts, attending to the precise details of explanations or descriptions. (MS-PS3-1), (MS-PS3-5)

RST.6-8.3 Follow precisely a multistep procedure when carrying out experiments, taking measurements, or performing technical tasks. (MS-PS3-3), (MS-PS3-4)

WHST.6-8.1 Write arguments focused on discipline content. (MS-PS3-5)

WHST.6-8.7 Conduct short research projects to answer a question (including a self-generated question), drawing on several sources and generating additional related, focused questions that allow for multiple avenues of exploration. (MS-PS3-3), (MS-PS3-4)

Tying It to *Common Core State Standards, Mathematics*

MP.2 Reason abstractly and quantitatively. (MS-PS3-1), (MS-PS3-4), (MS-PS3-5)

6.RP.A.1 Understand the concept of ratio and use ratio language to describe a ratio relationship between two quantities. (MS-PS3-1), (MS-PS3-5)

7.RP.A.2 Recognize and represent proportional relationships between quantities. (MS-PS3-1), (MS-PS3-5)

8.F.A.3 Interpret the equation $y = mx + b$ as defining a linear function whose graph is a straight line; give examples of functions that are not linear. (MS-PS3-1), (MS-PS3-5)

6.SP.B.5 Summarize numerical data sets in relation to their context. (MS-PS3-4)

Misconceptions

Some of the common misconceptions around this topic include the notions that heat and temperature are the same thing; that heat travels from cold to hot; that warm clothes heat the body; and that layering insulators on an object will heat it up.

To research these misconceptions, we suggest using the free Science Object from the NSTA Learning Center entitled "Thermal Energy, Heat, and Temperature and Energy Transformations." Select "middle school" and "free resources."

Dr. Philip Sadler and others associated with the Harvard-Smithsonian Center for Astrophysics have produced a series of videos and resources for science and mathematics educators regarding misconceptions and how to work with students to help them break through their misconceptions. You can find "Minds of Our Own" and other free online resources and video resources at *www.learner.org/resources/series26.html*.

We recommend investigating Page Keeley's series, *Uncovering Student Ideas in Science* (*http://uncoveringstudentideas.org*).

Objectives

By the end of this activity, students will have demonstrated the ability to:

- Use given materials to collect data on temperatures to determine the best insulator.

- Design, construct, and test a shade structure to determine which model provides the coolest temperature.

Academic Language

Conduction, convection, radiation, heat transfer, insulator, conductor, heat, temperature, thermal energy

Focus Question(s) (Scientific Inquiry)

- How might the type and/or color of a material affect planning a shade structure?
- In what way is thermal energy transferred to objects?
- Which materials are conductors and insulators of heat?
- How can you collect data to evaluate the effectiveness of a shade structure?

Framing the Design Problem(s) (Engineering Practice)

Design a shade structure that provides the lowest possible temperature under testing conditions.

Refer to Figure 4.3 "The Engineering Design Process" (p. 45) for more information about how to frame the design problem for this activity.

Teacher Background

This activity focuses on the idea of thermal energy and that heat transfer occurs in three ways: by radiation, conduction, and convection. To enhance teacher content knowledge of this topic, we suggest teachers research various internet resources including the NSTA Learning Center (see the Misconceptions section). Also, there are many good YouTube videos, Khan Academy videos, and TEDed resources to review content knowledge associated with thermal energy and heat transfer.

In this activity, students complete an initial investigation to explore the effect of color on temperature by recording data with thermometers on colored cans. They will have access to various materials to test them as possible resources to make their shade structure.

Part of the learning experience is letting students design how they will collect and record their data. This may vary per group, which makes for good discussion and argumentation. Writing up their results and evidence for argumentation directly correlates to *Common Core State Standards, ELA*.

Preparation and Management

Gather materials (listed below).

Prep Time

60 minutes to gather the materials and paint the cans

Teaching Time

Three or four 45-minute class periods

Materials

- Thermometers (low tech or high tech)
- 8 sets of soup cans painted various colors (red, white, black, blue, yellow, green)
- Miscellaneous materials to test and use for shade structures: wood strips, aluminum foil, batting, felt, fabric of various weights, plants, Styrofoam, cardboard, dirt, wire, string, masking tape (Invite students to bring items in as well.)
- Don't forget safety splash goggles, aprons and gloves.

Note: We've seen a couple of variations on this list. In Arizona, radiant heat from hot pavement adds a variable to the activity, depending on the time of year, so teachers sometimes used construction paper of various colors folded into envelopes that can easily be stored or recycled. (48 soup cans require a good chunk of storage space!)

5E Instructional Model

Engage

Ask students, "What do a rattlesnake, a kangaroo rat, and a teenager have in common?" Allow time for students to discuss in their table groups. Call for volunteers to share their guesses. After three guesses, if nobody lands on the right answer, tell them that all three species go to extraordinary measures to stay cool during the summer.

Preassessment

Tell the students that they will be learning the science of "being cool" (staying cool), but first they need to do a pre-assessment. Have students write the following academic language terms in their science notebooks: *thermal energy, convection, conduction, radiation, heat transfer, insulator, conductor, heat, temperature.*

Ask the students to use these words to create a concept map in their notebooks showing their understanding of the words. Look at their notebooks to assess their prior knowledge. Students who demonstrate an understanding of the concept of heat transfer should be pulled aside to discuss their understanding, and then given the option of working on a more challenging activity.

For preassessment, you can also query the class to gauge their prior knowledge:

- How is heat transferred?
- In what way do color and the type of material affect heat transfer?

These questions may be given as writing prompts and reviewed to see how and what students think about heat transfer and whether they use evidence and detail to support their arguments. This will help you adjust your lessons based on student needs and ability levels.

Explore

Tell the students that eventually they will design something using a set of materials, and that they'll need to measure the temperatures in spaces directly above and below the various materials. (Don't tell them anything else.) Talk about safety measures:

1. Use materials appropriately.
2. Be careful with glass thermometers.
3. Use caution when handling soup cans—they often have sharp edges that can cut skin.

Let the students know that they will be testing the temperature as it is affected by various colors and materials (the variables).

Place students in groups of four with access to thermometers and a variety of colored cans and building materials, and let them know they'll need to collect data on the temperature of various colored cans and building materials. Have students design a way to record their data in their notebooks. This is part of the learning process, as opposed to the teacher explicitly showing students how to record their data.

After the students collect and record their data, have them individually write their arguments defending the results that they found. This is important for clarifying what the students did as well as in meeting ELA standards. The data from each group should be put on a whiteboard or chart paper, and the group should prepare to present to the class for all to compare and contrast. This is where students will note how other students recorded their data. The teacher can then elicit ideas from the students about which data collection method seemed to be the most helpful.

Allow the students to have access to the various materials and invite them to put the various materials around the cans and collect temperature data, again to test for changes. They should devise a chart to record the data collected. Have students put their new data on a whiteboard and conduct a class discussion.

Explain

As students stand in a circle with whiteboards displayed, discuss the results of the data collection. How did students collect and record their data? Which color had the highest and lowest temperatures? What material and color combination had the highest and lowest temperatures?

As the discussion proceeds, introduce the activity's academic language: *conduction, convection, radiation, heat transfer, insulator, heat,* and *temperature,* when appropriate.

Here are possible prompts for argumentation:

1. What factors in the data collection method affect the results?
2. Which methods of recording data seem to be most appropriate?
3. How does ratio play a part in explaining the results of the inquiry?
4. What evidence is there for the type of energy being transferred?
5. What are some "real life" examples that fit the conduction (touching a spoon in a hot cup of liquid), convection (cooking pasta), and radiation (standing by a fire or open oven door) explanations of heat transfer?

Elaborate (Engineering Design Challenge)

Take students to a place on campus where students gather but which has no shade cover. Challenge them to look at their data collected over the past two days and have them design and construct a model of a shade cover that could be placed at a location on campus that would allow students to congregate and remain cool and comfortable.

After all the structures have been designed, have each group discuss their solution, incorporating the academic language from the previous day.

Evaluate

Have students create a concept map using the academic language (same exercise as in preassessment): *conduction, convection, radiation, heat transfer, insulator, heat, temperature.*

Students should also draw and label a model of their shade structure and explain why this design would be optimal.

Discussion and Argumentation

As students unveil their plans for their shade structures, they must use the data they collected to support their arguments for the color and materials used for their structure. Challenge the audience members to consider variables that might have impacted the data or could otherwise alter the outcomes.

Differentiation

1. **Broader Access Activity:** Structure groups to incorporate students of varied abilities. Think about limiting or adding supplies as needed. Since this activity is very hands-on, students who do not excel in written or bookwork types of activities might find success. Look for opportunities for them to become leaders.

2. **Extension Activity:** Allow students to investigate and experiment with passive solar or adding different fabrics or other materials to the cans that can be used to further block heat transfer.

3. **Modified Assessment:** Allow students to share their understanding of the concept orally with you. Some students may need to draw a picture or hold the objects to explain their understanding.

4. **Challenge Assessment:** Ask students to consider ways to improve their design and provide alternatives, including how they would test each one for the intended purpose. Students should prepare written and oral arguments for their alternate designs.

Note: These are examples as illustrations of possible differentiation options. The actual adaptations you create will depend on the results of your preassessment and ongoing formative assessments of individual students.

Activity 3: Bridge Buster
Engineering: Structural Design

In our daily lives, we rely on many different structures to provide us with shelter and access to new locations. Roads, bridges, and pipelines are just a few examples of "critical infrastructure" that we often take for granted.

This activity places students in the roles of scientists and engineers as they learn about the forces of gravity and the interaction between design and the strength-to-mass ratio, or as it is often termed, the strength-to-cost ratio. In this activity, students can use literature and internet resources to design and build a toothpick bridge. Required baseline characteristics are as follows:

- The bridge must support the mass of a toy car selected by the teacher. The bridge must be designed such that the car can be pushed and roll the entire length of the bridge.
- The bridge supports will be placed such that there is an unsupported span of 50 cm. (Depending on the cost of materials and time for this activity, you may adjust the length of the bridge to best fit your needs. We suggest a minimum length of 25 cm.)
- The only materials available for the construction include toothpicks and white glue (e.g., Elmer's Glue). Another design variation can be the use of a single thickness of standard notebook paper trimmed to fit the "pavement surface" of the bridge. This paper surface will aid the smooth rolling of the car during the testing phase.

Safety First

In every activity, we remind you to be certain that you understand the potential risks involved and are confident you can ensure your students' safety. Before attempting any of these activities in class, we recommend completing them yourself, and optimally with a teaching partner.

STEM

Bridge Buster encourages students to investigate and develop their scientific knowledge of structures, forces, and motion. Then, students use their scientific knowledge to design a piece of technology: the maximized-efficiency toothpick bridge.

This activity requires students to enlist their innovation and creativity and practice making claims based on evidence. Then, students must take their evidence-based conclusions to design a device according to specific standards.

This is a great activity to familiarize students with a design process. They start with researching readily available examples and internet resources and use that knowledge to actually design, build, and test a device according to its design specifications.

There's also some peripheral learning in this activity too. We've found working with students that there's often a significant difference between what we perceive we can do in a project based on instructions on the internet, and what it really takes to accomplish an engineering challenge. This can lead to good impromptu discussions during or after the activity.

Tying It to *NGSS*

Define the criteria and constraints of a design problem with sufficient precision to ensure a successful solution, taking into account relevant scientific principles and potential impacts on people and the natural environment that may limit possible solutions. (MS-ETS1-1)

Evaluate competing design solutions using a systematic process to determine how well they meet the criteria and constraints of the problem. (MS-ETS1-2)

Analyze data from tests to determine similarities and differences among several design solutions to identify the best characteristics of each that can be combined into a new solution to better meet the criteria for success. (MS-ETS1-3)

Develop a model to generate data for iterative testing and modification of a proposed object, tool, or process such that an optimal design can be achieved. (MS-ETS1-4)

Science and Engineering Practices

Asking Questions and Defining Problems

Define a design problem that can be solved through the development of an object, tool, process, or system and includes multiple criteria and constraints, including scientific knowledge that may limit possible solutions. (MS-ETS1-1)

Asking questions and defining problems in grades 6–8 builds on grades K–5 experiences and progresses to specifying relationships between variables, clarifying arguments and models.

Ask questions to identify and clarify evidence of an argument. (MS-ESS3-5)

Developing and Using Models

Develop a model to generate data to test ideas about designed systems, including those representing inputs and outputs. (MS-ETS1-4)

Plan for safety by determining which engineering controls, safety procedures and personal protective equipment will be needed.

Analyzing and Interpreting Data

Analyze and interpret data to determine similarities and differences in findings. (MS-ETS1-3)

Engaging in Argument From Evidence

Evaluate competing design solutions based on jointly developed and agreed-upon design criteria. (MS-ETS1-2)

Disciplinary Core Ideas

ETS1.A: Defining and Delimiting Engineering Problems

The more precisely a design task's criteria and constraints can be defined, the more likely it is that the designed solution will be successful. Specification of constraints includes consideration of scientific principles and other relevant knowledge that are likely to limit possible solutions. (MS-ETS1-1)

ETS1.B: Developing Possible Solutions

A solution needs to be tested, and then modified on the basis of the test results, in order to improve it. (MS-ETS1-4)

There are systematic processes for evaluating solutions with respect to how well they meet the criteria and constraints of a problem. (MS-ETS1-2), (MS-ETS1-3)

Sometimes parts of different solutions can be combined to create a solution that is better than any of its predecessors. (MS-ETS1-3)

ETS1.C: Optimizing the Design Solution

Although one design may not perform the best across all tests, identifying the characteristics of the design that performed the best in each test can provide useful information for the redesign process—that is, some of those characteristics may be incorporated into the new design. (MS-ETS1-3)

The iterative process of testing the most promising solutions and modifying what is proposed on the basis of the test results leads to greater refinement and ultimately to an optimal solution. (MS-ETS1-4)

Crosscutting Concepts

Influence of Science, Engineering, and Technology on Society and the Natural World

All human activity draws on natural resources and has both short and long-term consequences, positive as well as negative, for the health of people and the natural environment. (MS-ETS1-1)

The uses of technologies and limitations on their use are driven by individual or societal needs, desires, and values; by the findings of scientific research; and by differences in such factors as climate, natural resources, and economic conditions. (MS-ETS1-1)

Tying It to *Common Core State Standards*, ELA

RST.6-8.7 Integrate quantitative or technical information expressed in words in a text with a version of that information expressed visually (e.g., in a flowchart, diagram, model, graph, or table). (MS-ETS1-3)

WHST.6-8.7 Conduct short research projects to answer a question (including a self-generated question), drawing on several sources and generating additional related, focused questions that allow for multiple avenues of exploration. (MS-ETS1-2)

WHST.6-8.8 Gather relevant information from multiple print and digital sources; assess the credibility of each source; and quote or paraphrase the data and conclusions of others while avoiding plagiarism and providing basic bibliographic information for sources. (MS-ETS1-1)

WHST.6-8.9 Draw evidence from informational texts to support analysis, reflection, and research. (MS-ETS1-2)

Tying It to *Common Core State Standards, Mathematics*

MP.2 Reason abstractly and quantitatively. (MS-ETS1-1), (MS-ETS1-2), (MS-ETS1-3), (MS-ETS1-4)

Common Misconceptions

There are many misconceptions about forces and material strength, as well as misconceptions regarding bridge design and strength.

One common misconception is that if an object "appears" to be stationary (not moving) there are no forces acting on the object.

First, it is important to understand that what we perceive is in reference to our own experience at that point in time. For example, I feel like I am stationary as I sit in this chair typing these words. However, I am actually sitting on a chair and experience the effect of gravity between myself and the Earth, as the Earth is moving through the solar system, galaxy, and universe.

Additionally, while bridges appear to be stationary and you may not "see" evidence of forces on bridges, when a car, truck, or other object rests on or moves across a bridge, there are forces affecting the bridge.

This misconception emerges in part because we build understanding from our experience. Our experience is from our perspective—what we actually see, feel, and can sense—instead of a broader understanding that the Earth is moving, for example.

If you search the internet, you will find many websites that provide a range of explanations and accurate representations regarding forces on objects that appear stationary. We would need many pages to accurately describe and explain the factual scientific information regarding forces and design principles used for building bridges, tall buildings, and many other engineered structures. Therefore, we suggest using various resources including YouTube, Khan Academy, TEDed *Bridge Design (and Destruction!) Part 1* and other TEDed videos and lessons, as well as the NSTA Learning Center to locate resources that can help you answer questions that will likely arise from your students.

Dr. Philip Sadler and others associated with the Harvard-Smithsonian Center for Astrophysics have produced a series of videos and resources for science and mathematics educators regarding misconceptions and how to work with students to help them break through their misconceptions. You can find "Minds of Our Own" and other free online resources and video resources at *www.learner.org/resources/series26.html*.

Another resource we recommend for this section in each activity is Page Keeley's series *Uncovering Student Ideas in Science*: *http://uncoveringstudentideas.org*.

(Remember that we repeat some of these resources in our activities—on purpose—for those of you who are not reading this book cover to cover.)

Objectives

By the end of this activity, students will have demonstrated the ability to

- develop a performance-based project to demonstrate understanding of principles associated with building bridges;
- acquire scientific knowledge regarding the design principles associated with force and load distribution in the design of a toothpick bridge;
- design, build, and test a toothpick bridge within design parameters; and
- engage in scientific argumentation based on empirical data.

Academic Language

Force, load, gravity, parameter, claims and evidence, argumentation, data, empirical knowledge, science, mass, technology

Focus Question(s) (Scientific Inquiry)

- Name [optional: draw on whiteboard] some examples of famous bridges.
- What would be tricky about building a bridge?
- What materials would be ideal to build the strongest bridge possible?
- What do you think is the strongest part of a bridge span? The weakest part?
- What factors determine the strength or durability of a bridge?

Framing the Design Problem(s) (Engineering Practice)

Ask the students to design, build, and test the lowest mass and strongest bridge with the materials provided. The winning design will maximize the strength-to-mass ratio by maintaining continuity with the greatest mass attached at the midpoint of the bridge structure.

This is a test that will result in the destruction of the toothpick bridge. Students should be informed of this prior to beginning the project. Also, you will need to measure the mass of the bridge prior to testing the strength to the point of failure/destruction.

Refer to Figure 4.3 "The Engineering Design Process" (p. 45) for more information about how to frame the design problem for this activity.

Teacher Background

Depending on your familiarity with the science behind bridge construction, Newton's laws, and forces, we recommend you review the resources listed in the "Misconceptions" section, and particularly the NSTA Learning Center's material (Newton's laws and forces).

The ideal learning scenario for this activity includes allowing students to develop explanations based their experience in building their bridge. Failures should be cast as learning opportunities and capitalized upon to improve the final product. Then, the teacher can help students agree on a common understanding based on the evidence collected as a class data set.

Here are some specific pieces of information that may be useful in preparing for this activity, depending on your background and experience (*Note:* in these examples, the word "strength" represents the greatest mass that was supported before the bridge breaks during a "strength test." We are not saying that strength equals mass, but that the word is representative for the greatest mass that was supported prior to the final testing phase of the bridge designs.):

- The ratio of strength to mass is expressed mathematically as maximum mass supported by the bridge/mass of the bridge = ratio.
- Strength is defined in this situation as the mass that is successfully supported by the bridge.
- For example, a bridge has mass of 150 g and supports 75 g stacked or suspended at the midpoint of the bridge. The ratio would be 75 g / 150 g = 0.5. For a second bridge with a mass of 300 g and supports 75 g stacked or suspended at the midpoint of the bridge. The ratio would be 75 g / 300 g = 0.25. In this example the first bridge has the maximized strength-to-mass ratio because 0.5 is greater than 0.25.

Preparation and Management

Gather the materials for the student workgroup kits. Each student work group will need the following:

- 1 box of wood toothpicks (More boxes of toothpicks may be used per group depending on the size and time you decided to dedicate to the lessons. Also, note that the more toothpicks used, the greater the mass of the bridge the effect on the ratio.)
- 1 bottle of white wood glue, such as Elmer's or another brand
- Safety glasses or goggles

Prep Time

1–3 hours to gather materials (depending on availability of materials)

20 minutes to construct the kits student teams will use for the *Engage* and *Explore* phases and for the Design Challenge

Teaching Time

Two to four teaching periods (generally 45 minutes per teaching period)

Materials

- Wood toothpicks
- Wood glue
- Balance to measure the mass of the bridge
- Items of various masses to use sequentially to strength test the student projects
- Safety glasses or goggles

5E Instructional Model

Engage

Option 1: Show the Tacoma Narrows Bridge video clip (available through various internet video archives, such as YouTube, Vimeo, and so on).

Option 2: Show any video clips of bridge-building accomplishments.

Questions for the *Engage* phase include the following:

- What do you see?
- How do you explain what you see?

Preassessment

Useful introductory questions to gauge student background knowledge include the following:

- How can you explain this phenomenon?
- What evidence do you have for your explanation?

These questions will help you assess prior knowledge and determine whether your students have any previous exposure to engineering or physical science, while probing their argumentation skills. Moving forward, you can shape instruction more effectively with a better idea of students' ability levels and needs.

Explore

Tell the students that they will be given a kit to design and build a bridge. In any engineering project, there are limitations and specifications regarding time for completion, costs, supplies, dimensions, and capacities. Give the students the specifications that you have decided on based on the level of your students and the class time available for this activity. See the activity instructions in the Framing the Design Problem and Teacher Background sections.

Strength testing note: We suggest you use a standard paint-stirring stick (usually free at most home improvement stores). Lay the stick flat at the midpoint of the toothpick bridge and suspend nylon cord from each end. Join the ends of the cord and place weights at the end of the cord suspended from the stick across the midpoint of the bridge. If the bridge holds for 30 seconds, add additional mass to approach a breaking point. This should be standard process for "fairness" in testing all bridges.

Explain

Each team of scientists should produce a whiteboard set of claims and evidence regarding their predictions for the design's anticipated results. These should be shared with the class. The teacher facilitates the discussion using data collected from the whiteboarding activity. After each group has explained its claims and data, the information should be written and published so individual groups can make comparisons with the class set of common claims and evidence. This should evoke questions about any variation between the class set of common claims and the claims each team made based on their evidence. (This is an excellent opportunity to implement formative assessment practices for clarifying and reteaching topics and information; this may also be necessary to improve the accuracy of what the students have learned.)

Elaborate

Design Challenge

Students will need to write a narrative description of their design elements that they present, conducting research using reliable internet sources, in preparation to develop and implement the bridge design they produced in the *Explore* phase. Also, in this section they should predict how they would redesign their next toothpick bridge to improve the original design's strength-to-mass ratio.

The students will need to provide a set of data to support their claims for their first bridge and for a redesigned second bridge. Additionally, students will need to provide a written elaboration of their strength-to-mass ratio. Included in their written elaboration should be the decisions they made (and their rationale for the decisions) to arrive at their final design. Elaborations must include mathematical representations as well as a narrative and labeled diagrams.

As noted in the Materials section, you will need to provide a balance and items of various masses to measure the strength of the bridges.

Evaluate

Students are evaluated based on the depth of understanding expressed in their *Exploration* as well as *Elaboration* phases. The team with the greatest strength-to-mass ratio won't necessarily earn the highest evaluation. The evaluation should also be based on the quality of detail and alignment between claims and evidence and the *Elaboration* information from the engineering design challenge.

Discussion and Argumentation

Students should be given multiple opportunities to voice and represent their claims and evidence and refine their claims based on class-pooled evidence. This dialogue may be facilitated by the teacher and should focus on the nature of scientific ways of knowing (systematic review of data and the processes used to collect the data). We are not in search of "one correct answer." Instead, we suggest the students stay focused on exploring and articulating what they "know" and the evidence they have. If the students need to go back to the activity to run additional tests, that should be encouraged. All of this, of course, depends on how much time you can dedicate to the activity.

If Time Allows

One of the many important opportunities provided in this activity is the chance to help students experience the analysis and redesign aspects expected in the *NGSS* domain of science and engineering practices.

This can be accomplished in less time than the first cycle of this lesson. Students already have a familiarity with the materials and processes. Taking an additional 45 minutes or more can enable surprising advances in student learning and understanding.

We suggest that you review the findings of the class, and allow one or more school days to pass to allow students to use out-of-school time (safely!) to explore options regarding improving their design and construct a new and improved toothpick bridge. This will not cost any additional instructional time, other than your standard safety reminders and instructions, and may allow students to improve on their bridge designs in novel ways. Then, you can have one or two 45-minute sessions for students to construct and analyze data from another cycle of investigation, starting with *Explore*, and then moving into *Explain*, *Elaborate*, and *Evaluate*. This is, again, as time permits.

Differentiation

1. **Broader Access Activity:** Based on data for each student, provide suggested readings and internet sites with reliable resources.

2. **Extension Activity:** Minimize the directedness of the activity; alternately, ask students to develop a fair test for determining the strength-to-mass ratio. This could include various levels of additional information. This additional complex analytical task would need to include an evaluation of the protocol and potential outcomes of the proposed fair tests.

3. **Modified Assessment:** Modify the assessment to focus on the verbal, written, or diagrammatic elaboration of the principles in this lesson.

4. **Challenge Assessment:** Allow students to present a comparison of the variables associated with the design and resource parameters. They could present their findings in multiple representations and provide a rationale for how they decided to present the information.

Note: These are examples as illustrations of possible differentiation options. The actual adaptations you create will depend on the results of your preassessment and ongoing formative assessments of individual students.

CHAPTER 11

Activity 4: An Ounce of Prevention
Earth Science: Natural Disasters and Engineering Technology

Some of the most severe weather in the past decade has wreaked havoc on coastal cities and "tornado alley" in the Midwest.

Engineers are hired by cities to be proactive, creative problem solvers when it comes to protecting cities from natural hazards. In this activity, students get to be part of this process by exploring the science of natural hazards, engineering viable solutions, and testing them on "sand" towns.

Safety First

In every activity, we remind you to be certain that you understand the potential risks involved and are confident you can ensure your students' safety. Before attempting any of these activities in class, we recommend completing them yourself, and optimally with a teaching partner.

STEM

This activity is a well-integrated STEM lesson. Students learn about the science of natural hazards by exploring them on the internet and working on a design solution to a problem. Students will use the free Google SketchUp program to design and revise structures or solutions to mitigate the damage natural hazards can cause. As students work through the design process, they will develop a model and the students will test their solution in a "real life" student-designed simulation. All the STEM skills will be in full swing in this lesson, which is a great engineering extension to MS-ESS3g.

Tying It to NGSS

Define the criteria and constraints of a design problem with sufficient precision to ensure a successful solution, taking into account relevant scientific principles and potential impacts on people and the natural environment that may limit possible solutions. (MS-ETS1-1)

Science and Engineering Practices

Asking Questions and Defining Problems

Asking questions and defining problems in grades 6–8 builds on grades K–5 experiences and progresses to specifying relationships between variables, clarifying arguments and models. Define a design problem that can be solved through the development of an object, tool, process, or system and includes multiple criteria and constraints, including scientific knowledge that may limit possible solutions. (MS-ETS1-1)

Disciplinary Core Ideas

ETS1.A: Defining and Delimiting Engineering Problems

The more precisely a design task's criteria and constraints can be defined, the more likely it is that the designed solution will be successful. Specification of constraints includes consideration of scientific principles and other relevant knowledge that are likely to limit possible solutions. (MS-ETS1-1)

Crosscutting Concepts

Influence of Science, Engineering, and Technology on Society and the Natural World

All human activity draws on natural resources and has both short- and long-term consequences, positive as well as negative, for the health of people and the natural environment. (MS-ETS11)

The uses of technologies and limitations on their use are driven by individual or societal needs, desires, and values; by the findings of scientific research; and by differences in such factors as climate, natural resources, and economic conditions. (MS-ETS1-1)

Tying It to *Common Core State Standards*, ELA

RST.6-8.1 Cite specific textual evidence to support analysis of science and technical texts. (MS-ETS1-1), (MS-ETS1-2), (MS-ETS1-3)

RST.6-8.7 Integrate quantitative or technical information expressed in words in a text with a version of that information expressed visually (e.g., in a flowchart, diagram, model, graph, or table). (MS-ETS1-3)

RST.6-8.9 Compare and contrast the information gained from experiments, simulations, video, or multimedia sources with that gained from reading a text on the same topic. (MS-ETS1-2), (MS-ETS1-3)

WHST.6-8.7 Conduct short research projects to answer a question (including a self-generated question), drawing on several sources and generating additional related, focused questions that allow for multiple avenues of exploration. (MS-ETS1-2)

WHST.6-8.8 Gather relevant information from multiple print and digital sources; assess the credibility of each source; and quote or paraphrase the data and conclusions of others while avoiding plagiarism and providing basic bibliographic information for sources. (MS-ETS1-1)

WHST.6-8.9 Draw evidence from informational texts to support analysis, reflection, and research. (MS-ETS1-2)

SL.8.5 Include multimedia components and visual displays in presentations to clarify claims and findings and emphasize salient points. (MS-ETS1-4)

Tying It to *Common Core State Standards, Mathematics*

MP.2 Reason abstractly and quantitatively. (MS-ETS1-1), (MS-ETS1-2), (MS-ETS1-3), (MS-ETS1-4)

7.EE.3 Solve multi-step real-life and mathematical problems posed with positive and negative rational numbers in any form (whole numbers, fractions, and decimals), using tools strategically. Apply properties of operations to calculate with numbers in any form; convert between forms as appropriate; and assess the reasonableness of answers using mental computation and estimation strategies. (MS-ETS1-1), (MS-ETS1-2), (MS-ETS1-3)

7.SP Develop a probability model and use it to find probabilities of events. Compare probabilities from a model to observed frequencies; if the agreement is not good, explain possible sources of the discrepancy. (MS-ETS1-4)

Misconceptions

It is important to note a couple of common misconceptions about this topic. First, it's a fairly common belief that Earth events taking place within the global environment are not interconnected. For instance, the notion that weather systems operate independently is untrue, yet people in the Midwest may assume that El Nino doesn't impact them. It's a common misconception that the atmosphere, hydrosphere, lithosphere, and biosphere are not interrelated. On some level, these knowledge claims have been politicized, too, as teachers in the middle grades should be aware.

We suggest using the free or nominal cost journal resources from NSTA's Learning Center, *http://learningcenter.nsta.org*. Select "middle school" and "natural disasters." There are other resources on the TedEd site that may be helpful also.

Once again, we point readers to the series of resources produced by Dr. Philip Sadler and others associated with the Harvard-Smithsonian Center for Astrophysics, regarding misconceptions and how to work with students to help them break through their misconceptions. You can find "Minds of Our Own" and other free online resources and video resources at *www.learner.org/resources/series26.html*.

We also recommend that teachers consult Page Keeley's series *Uncovering Student Ideas in Science* (*http://uncoveringstudentideas.org*).

Objectives

By the end of this activity, students will have demonstrated the ability to

- research a natural hazard then make a chart comparing and contrasting the various hazards;
- use given materials to design and construct structures to mitigate the effects of a natural hazard; and
- test their design on a model "sand" town.

Academic Language

Engineering and design process, avalanche, landslide, coastal erosion, hurricane, tornado, flood, drought, forest fire, earthquake

Focus Question(s) (Scientific Inquiry)

- Are natural hazards preventable? Why or why not?
- What are the limits on designs to protect people from natural hazards?
- How can devices or structures be designed to protect against the different strength levels of a natural hazard, for example, a Category 1 hurricane compared to a Category 5 hurricane, or a 5.0 vs. a 7.5 earthquake?
- In what ways do the long-term effects of natural disasters affect a population or ecosystem?
- How can models be used to test protective designs?

Framing the Design Problem(s)

Students will design a structure or technological device to mitigate the effects of a natural hazard. Teachers should plan for safety by determining which engineering controls, safety procedures, and personal protective equipment will be needed.

Refer to Figure 4.3 "The Engineering Design Process" (p. 45) for more information about how to frame the design problem for this activity.

Teacher Background

While we can't prevent natural disasters, engineering science offers potential to mitigate their impact on humans. A more disaster-resilient society will emerge through three major principles:

- Anticipate and assess risk; do not simply react to disasters.
- Focus on mitigation that builds resilience.
- Implement warning and information dissemination systems that allow society to bring its resilience into play.

Three areas students might consider and research in exploring how to mitigate the impact of natural disasters are integrated warning systems, enhanced building design and construction, and watershed planning and management. However, as much as possible, teachers should let students figure this out on their own through their research efforts.

Teacher content knowledge can be enhanced through the NSTA Learning Center (see the Misconceptions section).

Google SketchUp is a free drawing tool that can be used by anyone and offers a wide range of opportunities for learning and drawing for levels from novice to expert. A range of training options are available at *www.sketchup.com* when you click on "Learn." The video tutorials under the "Learn" tab are an excellent place to start. The direct URL for the video tutorials for SketchUp is *www.sketchup.com/learn/videos?playlist=58.*

Preparation and Management

Depending on time available in class, have the students set up a "natural hazard scene" for each table group. (If you don't have the class time, you can prep the scenes yourself.) To make each diorama, take the lid of a copy paper box (or one of a similar size), cover it with a large plastic garbage bag and fill the covered lid with sand spread evenly on the bottom of the lid, 5 cm deep. (Ideally we'd use large, shallow plastic tubs that are reusable and less messy, but more costly initially.) Set up a scene similar to a town, with small items to represent houses, trees (small sprigs off of a bush will work), and other items that might be in a town. Be sure to form mountains and rivers out of the sand in places. You will need to reuse the items for other classes. (Alternatively, you can stage the diorama as if the disaster already happened and have students speculate about what took place.)

Prep Time

1 hour to gather materials; an additional 60–90 minutes if you opt to set up the disaster scenes in the box lids yourself. As mentioned, download the free Google SketchUp program (3D modeling) on available computers; review the program at *www.sketchup.com*.

Teaching Time

45 minutes during three class periods; more depending on extent of research project

Materials and Costs

- Various building materials: scraps of wood, aluminum foil, linoleum scraps, cardboard, and so on
- Safety goggles, aprons, face masks, and gloves
- Blow dryers (We borrow these from our art teachers; be sure to plug them into only GFI- or GFCI-protected wall receptacles to prevent shocks.)
- Containers of water
- Large, shallow plastic tubs (more costly initially, but reusable) or cardboard lids and garbage bags
- Sand for dioramas where students will test their engineered designs

5E Instructional Model

Engage

Have students sit in table groups and brainstorm a list of natural hazards and the effects of the hazards (e.g., tornado; trees blown down, houses destroyed). Have student groups report out as the teacher compiles a list on the board.

Direct students to study the scene of the towns created in the box lids. Tell them that a series of natural hazards are about to happen to the town. Have them put on goggles and watch closely to see what happens as they use a blow dryer (on the "no heat" setting) to simulate a tornado, sprinkle water on the town to replicate a hurricane, and pour water down the mountains and rivers to simulate flooding. The box can be bent to create an earthquake. (Depending on the class and circumstances, you might demonstrate this with them first, or do it simultaneously.) In a notebook, each student should write down what was happening in the town—the causes and effects of the natural hazard.

Preassessment

Questions you can use to probe students' prior knowledge include the following:

- "How do natural disasters affect a human's habitat?"
- "What are some ways to keep natural disasters from destroying property?"

These questions may be given as writing prompts and reviewed to determine whether your students have any exposure to Earth science in general and natural disasters specifically. You'll also get to see their argumentation skill level. This will help you shape the activity and subsequent lessons around student needs and abilities.

Safety Note

With some classes, it's more realistic to have each table group simulate a single type of natural disaster, and adjust the rest of the activity accordingly. If you opt for each table group to do multiple "disasters," be vigilant about the potential safety issues with the hair dryer and water, not to mention making sure the students use the "no heat" setting on the hair dryers. Have students immediately wipe up any liquid spilled on the floor to prevent slip-and-fall hazards. Have students wash their hands with soap and water after completing this activity.

Explore

Let the groups choose (or creatively assign) one natural disaster to research. Sometimes we "reward" the group with the most historically destructive hurricane, flood, and so on, by declaring them the experts assigned to study that type of disaster. Spread out the disaster types to minimize redundancy.

Have students work in table groups to investigate their disasters to find out the science behind what causes the event. They should have access to the internet or other resources in the classroom to research their hazard. Students can simply put their research findings on a small whiteboard and share with the class, or do an actual research paper on one of the natural hazards. After researching their natural hazard, students should be given time to brainstorm a device or structure they could create. They should be prepared to share their ideas.

Explain

Have each group present its natural hazard research information to the class. Other class members should draw a chart or table in their notebooks and write down key notes about each hazard. Each group should share its engineered design and tell the anticipated results of their test or their idea if it is a technological design.

The teacher's role is to act as a facilitator for this conversation, making sure that all students are completing their charts and encouraging them to ask questions of presenting groups.

Elaborate

Design Challenge

As an extension of their research, students need to work through the engineering and design process by designing a technological device or structure that will help mitigate the damage their natural hazard causes. These can be done using Google SketchUp, or it could be done in their lids or tubs. Allow access to various building materials, if desired. If a group is designing a technological device, the members need to have a detailed drawing with full explanations done on Google SketchUp.

Students should be given time to test their prototypes on their sand towns. They should collect and organize their data in their notebooks in preparation for the class discussion (and this can also be put on a whiteboard).

Each group will then share its engineered design with the class. Groups may opt out to provide the results of their tests or details of their ideas if they have technological designs that are impractical to prototype in class.

Evaluate

Based on their own experiences as well as the class presentations, students should write about at least three different natural hazards, contrasting their differences and their impacts on human populations, as well as describing the engineered design they made and the test they conducted. Students should be encouraged to reflect on their designs, listing pros and cons of their projects and what they would do next to make improvements or try a new design.

Discussion and Argumentation

As groups present their research, other groups should be encouraged to ask questions to probe for understanding and ask questions about the design process their peers used.

Differentiation

1. **Broader Access Activity:** This activity is very much a team activity and benefits from allowing the strengths of all students to be used. Encourage each group member to think about his or her area of expertise and be actively involved in the Design Challenge. Be sure all members of the team share their process as part of their team's oral presentation.

2. **Extension Activity:** Challenge students by having them execute the next phase of the Design Challenge, designing and/or developing an improved solution.

3. **Modified Assessment:** Students can verbally explain or create a drawing or provide some other nonwritten representation of the process they went through during the Design Challenge.

4. **Challenge Assessment:** Students can analyze the feasibility of the solutions provided by each of the various groups.

Note: These are examples as illustrations of possible differentiation options. The actual adaptations you create will depend on the results of your pre-assessment and ongoing formative assessments of individual students.

Activity 5: Some Striped Seeds Seem Similar!

Biology: Genetic Variation

For almost all species, there is a range of variation among the individuals within a population of that species. Some of these variations affect the survivability of individuals, who then become vulnerable and may not endure. Some variations are affected by environmental factors. Some variations have no effect on survivability, or may not be affected by changes in the environment.

However, there is a pattern in nature that appears almost every time we measure or count individuals of a species based on a *specific characteristic*.

This activity puts students in the role of scientists as they learn about statistics and genetic variation within a species and use the statistical understanding to describe frequencies of specific variations within a species. In this activity, students will focus on the discovery of statistical patterns that emerge from the data.

Safety First

In every activity, we remind you to be certain that you understand the potential risks involved and are confident you can ensure your students' safety. Before attempting any of these activities in class, we recommend completing them yourself, and optimally with a teaching partner.

STEM

This activity capitalizes on the patterns in nature and how close observations of natural objects can reveal a new level of appreciation for mathematical thinking as it is revealed in science.

This activity also introduces students to (or reinforces) a process in which they experience how *scientific knowledge* is developed and tested. Then, students use the scientific knowledge to *predict and test* other variations among members of another species.

Tying it to NGSS

Construct an explanation based on evidence that describes how genetic variations of traits in a population increase some individuals' probability of surviving and reproducing in a specific environment. (MS-LS4-4)

[Clarification Statement: Emphasis is on using simple probability statements and proportional reasoning to construct explanations.]

CHAPTER 12

Science and Engineering Practices

Analyzing and Interpreting Data

Analyze displays of data to identify linear and nonlinear relationships. (MS-LS4-3)

Analyze and interpret data to determine similarities and differences in findings. (MS-LS4-1)

Using Mathematics and Computational Thinking

Use mathematical representations to support scientific conclusions and design solutions. (MS-LS4-6)

Constructing Explanations and Designing Solutions

Apply scientific ideas to construct an explanation for real-world phenomena, examples, or events. (MS-LS4-2)

Construct an explanation that includes qualitative or quantitative relationships between variables that describe phenomena. (MS-LS4-4)

Plan for safety by determining which engineering controls, safety procedures and personal protective equipment will be needed.

Obtaining, Evaluating, and Communicating Information

Gather, read, and synthesize information from multiple appropriate sources and assess the credibility, accuracy, and possible bias of each publication and methods used, and describe how they are supported or not supported by evidence. (MS-LS4-5)

Disciplinary Core Ideas

LS4.B: Natural Selection

Natural selection leads to the predominance of certain traits in a population, and the suppression of others. (MS-LS4-4)

Crosscutting Concepts

Patterns

Graphs, charts, and images can be used to identify patterns in data. (MS-LS4-3)

Connections to Nature of Science

Science assumes that objects and events in natural systems occur in consistent patterns that are understandable through measurement and observation. (MS-LS4-1), (MS-LS4-2)

Tying It to *Common Core State Standards*, ELA

RST.6-8.1 Cite specific textual evidence to support analysis of science and technical texts, attending to the precise details of explanations or descriptions. (MS-LS4-4)

RST.6-8.9 Compare and contrast the information gained from experiments, simulations, video, or multimedia sources with that gained from reading a text on the same topic. (MS-LS4-3), (MS-LS4-4)

WHST.6-8.2 Write informative/explanatory texts to examine a topic and convey ideas, concepts, and information through the selection, organization, and analysis of relevant content. (MS-LS4-2), (MS-LS4-4)

SL.8.4 Present claims and findings, emphasizing salient points in a focused, coherent manner with relevant evidence, sound valid reasoning, and well-chosen details; use appropriate eye contact, adequate volume, and clear pronunciation. (MS-LS4-2), (MS-LS4-4)

Tying It to *Common Core State Standards, Mathematics*

MP.4 Model with mathematics. (MS-LS4-6)

6.RP.A.1 Understand the concept of a ratio and use ratio language to describe a ratio relationship between two quantities. (MS-LS4-4), (MS-LS4-6)

6.SP.B.5 Summarize numerical data sets in relation to their context. (MS-LS4-4), (MS-LS4-6)

7.RP.A.2 Recognize and represent proportional relationships between quantities. (MS-LS4-4), (MS-LS4-6)

Common Misconceptions

There are numerous misconceptions about genetics and variation among individuals of a species. One common misconception is that individual organisms are identical if they are the same species. Another common misconception is that visible differences between individuals indicate that they are not the same species.

Textbooks and other teacher resources perpetuate these misconceptions when they provide only a superficial explanation without paying attention to how the authors use technical vocabulary. Additional sources of misconceptions are the result of the application of nonscientific beliefs to influence the interpretation of scientific data.

If you search the internet, you will find many websites that provide a range of explanations about genetic variation. However, exercise caution when evaluating these internet sources. NSTA has excellent resources on the topic of evolution and the principles associated with it; evolution is, to say the least, a sensitive topic that deserves careful planning to present effectively in some school communities.

We do not have the space to go into great detail in this section; therefore, we suggest readers seek additional information from the NSTA website (*www.nsta.org*) and the NSTA Learning Center (*learningcenter.nsta.org*; search for genetic variation or genetics and heredity).

For more on helping students break through scientific misconceptions, consider the resources developed by Dr. Philip Sadler and others associated with the Harvard-Smithsonian Center for Astrophysics: see "Minds of Our Own" and other free online resources and video resources at *www.learner.org/resources/series26.html*.

We also urge readers to consider Page Keeley's series about identifying the prior knowledge your students bring to your classroom: see *Uncovering Student Ideas in Science* (*http://uncoveringstudentideas.org*).

Objectives

By the end of this activity, students will have demonstrated the ability to

- collect, analyze, and pool data;
- look closely at biological objects (sunflower seeds), and carefully record data regarding patterns among the individuals of a species;
- analyze data and use data tables and graphs to describe patterns in nature (the variation in the number of stripes on a sunflower seeds);
- follow procedures consistently; and
- work collaboratively to process large sets of observations.

Academic Language

Trend, bell curve, natural distribution, visual physical graph, biostatistics, range, domain, mean, median, variance, seed case, magnifying glass

Focus Question(s) (Scientific Inquiry)

- Are all sunflower seeds the same?
- How do individual sunflowers seeds vary?
- How much variation is there among individuals of the same species?
- How do scientists and engineers collect large data sets?
- What advantages do scientists and engineers have when they use large data sets for analysis?
- When do you know you have enough data to have a high degree of confidence to make a "claim based on evidence"?

Framing the Design Problem(s) (Engineering Practice)

Nonnative invasive species are destroying habitat and biodiversity. Biodiversity is a measure of the number of different species in a defined geographic area. The health and stability of an ecosystem is demonstrated by the biodiversity in that ecosystem.

For example, zebra mussels are mollusks that are destroying fishing, recreational boating, and shipping in the Great Lakes. This costs taxpayers millions of dollars each year to combat the invasive mussels.

One option for a design problem might look like this: "Based on what you have learned about variation among a species in this activity, propose some investigations that will provide you with the information you could use to develop a device, process, or regulation to reduce the number of new zebra mollusks that are introduced into the Great Lakes."

Refer to Figure 4.3 "The Engineering Design Process" (p. 45) for more information about how to frame the design problem for this activity.

Teacher Background

For this activity, students are going to count the number of white stripes on each striped sunflower seed. Students need to keep track of the number of seeds with each number of stripes and sort the seeds into piles based on the number of stripes on the sunflower seeds. After students have completed counting and sorting their sunflower seeds, they will

- add their data to the class data on the whiteboard, and
- bring their piles of seeds sorted by number of stripes and add each of their piles to the appropriate piles (cylinders; see p. 144) at the front of the room.

For this activity, you'll need to collect 16 graduated cylinders or other transparent containers and arrange them across a table in order of number of stripes (from 0 stripes to 15 stripes). Students will collect data in teams of two, and then pool their data as a class.

One way to easily collect the class data is to make a chart on the whiteboard such as the Class Data Chart (see sample data tables, below). As students complete their data collection, they will record their data in their lab notebook data tables as well as the Class Data Chart on the board. For additional impact, teachers may collect the class data as a visual physical graph—either simple piles of seeds, or in a series of transparent cylinders. (An additional variation in middle schools where teachers change students every class period. Teachers can collect data the seeds for each class in plastic bags, and combine multiple class periods of data to demonstrate the effects of increasing sample size on the shape of the physical graph of data.)

Individual Group Data Table

Number of Stripes	0	1	2	3	4	5	6	7	8	9	10	11	12	13	14	15
Number of seeds																

Class Data Table

Number of Stripes	0	1	2	3	4	5	6	7	8	9	10	11	12	13	14	15
Seed counts from Group 1																
Seed counts from Group 2																
Seed counts from Group 3																
Seed counts from Group 4																
Seed counts from Group 5																
Seed counts from Group 6																
Seed counts from Group 7																
Seed counts from Group 8																
Seed counts from Group 9																
Seed counts from Group 10																
Seed counts from Group 11																
Sum of seed counts																

We suggest teachers research various internet resources (including the NSTA Learning Center's information on genetics; there also are many good YouTube videos, Khan Academy videos, and TEDed resources) to review content knowledge associated with genetic variation.

Preparation and Management

You'll need to gather the materials for the student work group kits. Each student work group will need the following:

- 1 resealable bag of ¾–1 cup of striped sunflower seeds (with an optional magnifying glass to see the stripes more clearly). (*Safety note:* Some seeds contain pesticides or herbicides. Remind students not to eat the seeds and also wash their hands with soap and water after completing the activity.)
- Safety glasses or goggles

Prep Time

60 minutes to fill the bags of sunflower seeds and prep the cylinders.

Also, schedule time for students to use computers for internet access.

Teaching Time

45–90 minutes

Materials

- Student access to computers to access websites or information printed from websites on invasive species (Zebra mussels or other important invasive specie in your area), statistics, and biostatistics.
- Whiteboard, markers, and eraser for each student pair (Large chart paper and felt pens can be used in place of whiteboards.)

5E Instructional Model

Engage

Show PowerPoint slides (or YouTube/TeacherTube videos) that show extremes of size variation of living organisms.

Ask students to share what they have heard about or experienced regarding the range of sizes of a particular animal or plant, and have students offer any ideas they have to explain the variation. (Always ask students to provide evidence or their thinking that led them to the idea that they expressed.)

Note: This section is not about being right or having the best answer. The *Engage* phase is all about getting students to think and involve themselves in the topic cognitively. This section also provides important information about student thinking and misconceptions.

Preassessment

Here are two questions you can use to gauge prior knowledge:

- How can you explain the variation among these observations?
- What evidence do you have for your explanation?

These questions may be given as writing prompts or in small-group discussions that you observe. You'll discover how and what students think about their observations (and you'll glimpse their argumentation skills). Your preassessment serves to help you shape the rest of the activity and subsequent lessons.

Explore

Tell students that they will work in teams of two to do the work of biostatisticians as we investigate the variability of a characteristic in a living organism (the number of white stripes on a sunflower seed).

Pass out the bags of sunflower seeds and have the students count and sort the sunflower seeds to determine the number of white stripes on the seeds. They will need to record their individual team data, record their data on the class data set, and add their piles of sorted seeds into the class visual physical graph setup at the front of the classroom.

Students should be cautioned to not eat the sunflower seeds as they have been handled by students in several different classes. If it's permitted at your school, you could reassure the students that they will be provided with some sunflower seeds to eat at the end of the lesson.

Explain

Each team of biostatisticians needs to record the class data and analyze their team data as it compares to the class data. Once the teams complete their analyses and comparisons, they will compose whiteboard explanations of the benefits and limitations of data based on the sample size of the data. (How does the number of data points affect the reliability and validity of claims based on the data?)

These should be shared with the class, and the teacher facilitates the discussion. (This is an excellent opportunity to implement formative assessment practices for clarifying and reteaching topics and information that may be needed to improve the accuracy of what the students have learned.)

After each group has defended its information, the information should be written up in an "official report" to be presented (published) for their classmates, family, or friends to compel them to increase their critical thinking about information presented in advertisements and other presentations.

Elaborate

Students should build on the understanding that there is variability among most individuals of a species, to develop a series of potential devices or systems/processes to reduce the introduction of invasive species (e.g., zebra mussels).

Evaluate

Students are evaluated based on the depth of understanding expressed in their *Exploration* as well as *Elaboration* phases. The teams with the most thorough explanation of the effect of sample size on reliability of claims based on evidence should be given the opportunity to reveal their correct understanding to their classmates. *Evaluation* should be based on the quality of their detail and the alignment between claims and evidence in the report to their classmates, family, and friends, and the *Elaboration* information from the engineering Design Challenge.

Discussion and Argumentation

Students should be given multiple opportunities to voice their claims and evidence. They should also be given the opportunity to refine their claims based on class-pooled evidence.

Discussions are generally facilitated by the teacher and should focus on the nature of scientific ways of knowing (systematic review of data and the processes used to collect the data). We are not searching for one correct answer. Instead, we suggest the students stay focused exploring and articulating what they "know" and the evidence they have. If the students need to go back to the activity to run additional tests, that should be encouraged. The scope of Discussion and Argumentation depends on the time you have available—there is a lot of potential for higher order thinking here.

You'll have the opportunity to discuss and debate the notion that "the more data we collect, the better," and to critically review questions such as, "Is there a point where you have enough data to make a claim, and there is diminishing return on the investment of time and energy to continue collecting data? If there is a point of diminishing returns, how do you know when you are at the point where you don't need to collect additional data?"

Differentiation

1. **Broader Access Activity:** Limit the number of seeds in some of the bags. By limiting the number of seeds, students who are easily distracted would have their attention focused only on the items necessary for a particular stage of the activity and have a reasonable expectation of completing the counting and sorting task around the same time that other teams are ready to record their data on the class data chart.

2. **Extension Activity:** Expand the options and have students explore the limitations of variability for seeds versus adult body size, or versus sunflower stalk height, or versus any other biological variable. Also, students could explore the role of genetics and environment on the variability of other types of organisms.

3. **Modified Assessment:** Modify the assessment to focus on the verbal, written, or diagrammatic elaboration of the principles in this lesson.

4. **Challenge Assessment:** Allow students to present a comparison of the effects of genetics and environment on variability among individuals of a species. They might present the data in multiple representations and provide a rationale for how they decided to represent the information.

Note: These are examples as illustrations of possible differentiation options. The actual adaptations you create will depend on the results of your preassessment and ongoing formative assessments of individual students.

Activity 6: Conserving the Drops of Drips

Earth Science: Water Conservation and Human Impact on Earth Systems

Faucets: Every school has them. Many are inefficient, and even if they're efficient, their *use* is often inefficient.

Students will think differently about faucets after this place-based investigation. This real-life activity gets students rolling up their sleeves, collecting data, doing research, and then designing and communicating solutions to a problem right at their own school. This is an immediate and engaging way for students to delve into place-based learning.

Safety First

In every activity, we remind you to be certain that you understand the potential risks involved and are confident you can ensure your students' safety. Before attempting any of these activities in class, we recommend completing them yourself, and optimally with a teaching partner.

STEM

Students will use science and math in calculating the usage and water flow from the faucets in their science lab or in a consumer and family science lab. Then they will use technology to research and design an engineering solution to the water usage problem in their school.

Tying It to *NGSS*

Apply scientific principles to design a method for monitoring and minimizing a human impact on the environment. (MS-ESS3-3)

[Clarification Statement: Examples of the design process include examining human environmental impacts, assessing the kinds of solutions that are feasible, and designing and evaluating solutions that could reduce that impact. Examples of human impacts can include water usage (such as the withdrawal of water from streams and aquifers or the construction of dams and levees), land usage (such as urban development, agriculture, or the removal of wetlands), and pollution (such as of the air, water, or land).

Science and Engineering Practices

Asking Questions and Defining Problems

Asking questions and defining problems in grades 6–8 builds on grades K–5 experiences and progresses to specifying relationships between variables, clarifying arguments and models. Ask questions to identify and clarify evidence of an argument. (MS-ESS3-5)

Constructing Explanations and Designing Solutions

Apply scientific principles to design an object, tool, or processor system. (MS-ESS3-3)

Plan for safety by determining which engineering controls, safety procedures and personal protective equipment will be needed.

Engaging in Argument from Evidence

Construct an oral and written argument supported by empirical evidence and scientific reasoning to support or refute an explanation or a model for a phenomenon or a solution to a problem. (MS-ESS3-4)

Disciplinary Core Ideas

ESS3.C: Human Impacts on Earth Systems

Human activities have significantly altered the biosphere, sometimes damaging or destroying natural habitats and causing the extinction of other species. But changes to Earth's environments can have different impacts (negative and positive) for different living things. (MS-ESS3-3)

Typically as human populations and per-capita consumption of natural resources increase, so do the negative impacts on Earth unless the activities and technologies involved are engineered otherwise. (MS-ESS3-3), (MS-ESS3-4)

ESS3.D: Global Climate Change

Human activities, such as the release of greenhouse gases from burning fossil fuels, are major factors in the current rise in Earth's mean surface temperature (global warming). Reducing the level of climate change and reducing human vulnerability to whatever climate changes do occur depend on the understanding of climate science, engineering capabilities, and other kinds of knowledge, such as understanding of human behavior and on applying that knowledge wisely in decisions and activities. (MS-ESS3-5)

Crosscutting Concepts

Cause and Effect

Relationships can be classified as causal or correlational, and correlation does not necessarily imply causation. (MS-ESS3-3)

Cause and effect relationships may be used to predict phenomena in natural or designed systems. (MS-ESS3-1), (MS-ESS3-4)

Stability and Change

Stability might be disturbed either by sudden events or gradual changes that accumulate over time. (MS-ESS3-5)

Tying It to *Common Core State Standards*, ELA

WHST.6-8.7 Conduct short research projects to answer a question (including a self-generated question), drawing on several sources and generating additional related, focused questions that allow for multiple avenues of exploration. (MS-ESS3-3)

WHST.6-8.8 Gather relevant information from multiple print and digital sources; assess the credibility of each source; and quote or paraphrase the data and conclusions of others while avoiding plagiarism and providing basic bibliographic information for sources. (MS-ESS3-3)

Tying It to *Common Core State Standards*, *Mathematics*

6.RP.A.1 Understand the concept of a ratio and use ratio language to describe a ratio relationship between two quantities. (MS-ESS3-3), (MS-ESS3-4)

7.RP.A.2, Recognize and represent proportional relationships between quantities. (MS-ESS3-3), (MS-ESS3-4)

6.EE.B.6 Use variables to represent numbers and write expressions when solving a real-world or mathematical problem; understand that a variable can represent an unknown number, or, depending on the purpose at hand, any number in a specified set. (MS-ESS3-1), (MS-ESS3-2), (MS-ESS3-3), (MS-ESS34), (MS-ESS3-5)

7.EE.B.4 Use variables to represent quantities in a real-world or mathematical problem and construct simple equations and inequalities to solve problems by reasoning about the quantities. (MS-ESS3-1), (MS-ESS3-2), (MS-ESS3-3), (MS-ESS3-4), (MS-ESS3-5)

Misconceptions

One of the biggest misconceptions relating to water conservation is that it is only a concern for arid states, or that water only needs to be conserved when there is a drought. Water conservation is a global imperative, though it's particularly difficult to impress that on the general population (much less pre-adolescents) when there's no immediate, urgent shortage pressing down.

For more on helping students break through scientific misconceptions, consider the resources developed by Dr. Philip Sadler and others associated with the Harvard-Smithsonian Center for Astrophysics: see "Minds of Our Own" and other free online resources and video resources at *www.learner.org/resources/series26.html*.

Regarding addressing misconceptions in each activity, regardless of subject matter, we also suggest readers consider Page Keeley's series about identifying the prior knowledge your students bring to your classroom: see *Uncovering Student Ideas in Science* (*http://uncoveringstudentideas.org*).

Objectives

By the end of this activity, students will have demonstrated the ability to

- collect data on faucet use and water flow;
- design solutions to improve functionality, correct problems and minimize future recurrences of the problems; and
- construct a presentation including argumentation from evidence regarding their solutions and communicate their recommendations to peers, teachers, and school administrators.

Academic Language

Environment, resources, conservation, human impact, water flow

Focus Question(s) (Scientific Inquiry)

- In what ways do humans impact resources on the Earth?
- What are the ramifications if humans are neglectful of Earth's resources?
- How can we identify campus areas that might have high water usage, and thus potential waste?
- What is the best way to collect data on water waste and usage?
- What solution designs would best solve water waste problems?
- How can we calculate the expense of wasted water?

- How can students share their data and recommendations with school administration? Who else needs to know?
- What are the channels to work through to get things changed at the school level?

Framing the Design Problem(s) (Engineering Practice)

The problem for this activity is straightforward: Students will evaluate water usage at their school, paying attention to inefficiencies and misuse, and design solutions to improve efficiency and water conservation.

Refer to Figure 4.3 "The Engineering Design Process" (p. 45) for more information about how to frame the design problem for this activity.

Teacher Background

Part of the process of any research project is letting students figure out where to find the information to research. Resist the temptation to give them a list of possible sources. Student research will undoubtedly turn up information about how to conserve water by changing habits and behaviors, fixing leaks, replacing worn-out fittings, and using water-saving devices like lavatory faucet regulators.

There are really two things going on here: usage by people and water flow from faucets. As the teacher, you can split the task by having some table groups research usage issues and others consider ways to improve water flow. (The EPA's WaterSense program is a good source for information about the latter topic.)

Students need to investigate the student-accessible water facilities around campus, measure water flow when faucets are on, and identify any leaking or malfunctioning plumbing. When they locate a leaky faucet or sink, they'll measure the water loss by capturing and timing the flow or drips, and then calculate the daily, weekly, and annual water loss. Students may suggest regulators to reduce flow in functional faucets and further conserve water. They can research water rates and determine the financial costs to the school, if possible.

After collecting data on how water is being used and wasted with school faucets, students will design solutions as simple as making signs or announcements about water conservation, recommending that the school purchase water regulators. Whatever their solutions, they need to discover them through their own research, and it's acceptable for students in different classes to arrive at similar results.

Finally, students will prepare a presentation to classmates, faculty, and administration. Teachers need to notify their administrators ahead of time so they'll be (a) available and hopefully (b) receptive to the student design recommendations.

In terms of content background for the teacher, in addition to abundant Internet sources, there are many journal articles for free or at nominal cost dealing with water conservation on NSTA's website (*www.nsta.org*).

Preparation and Management

Schedule time for students to use computers for internet access. If students lack convenient internet access, make copies of articles relating to water conservation and control of water usage. Gather materials for data collection (see below).

Prep Time

30 minutes to collect materials

Teaching Time

3 teaching periods, or about 120 minutes

Materials and Costs

Have the following materials available for students to use in their data collection, letting them decide what to use and how to collect their data:

- Stopwatches
- Containers of various sizes to hold water
- Measuring cups
- Internet access
- Safety glasses or goggles if students are working in science lab or other school space where hazards are located (e.g. glassware, hazardous chemicals, and so on).

5E Instructional Model

Engage

Hold up a glass of water and ask students what they know about the connection between water and life (human and otherwise) on Earth; specifically, find out what they know about conserving water and whether they think it's an issue for them. Ask the students if they can identify any places at school where water might be wasted. (They will name many areas, but guide the focus back to the faucets in the science lab, consumer sciences lab, or other student-accessible and safely occupied places in the school.)

Preassessment

Questions you can use to gauge students' prior knowledge include the following:

- In what ways does wasting water affect humans?
- What specific solutions are there for conserving water?

These questions may be given as writing or small-group discussion prompts to evaluate student prior knowledge about water conservation. You'll also be able to glimpse their argumentation skills. Through pre-assessment, you can shape the activity and subsequent lessons on this topic based on student needs and ability levels.

Explore

With students seated in table groups of four, tell them that they will become super sleuths as they collect data on water usage and problems with the school water system itself. Let students know that each group will have access to a stopwatch, containers, and a measuring cup. Each group will need to devise a plan detailing how the members are going to measure the water flow of a faucet (assign different faucets for each group) or calculate the rate of water wasted in a sink through careless hand washing habits, for example. Have the students work through the Engineering Design process described in Fig. 4.3 on page 45. Each group member needs to use a science notebook to record their step-by-step plan for collecting data. After all members of the group have their plans recorded, they need to gather their materials and test their plan. (This may take a few tries for them to figure out, but let them wrestle with the outcomes.)

As students record their data in their science notebooks, make sure they're using the math skills in the standards listed above. Their research, according to the ELA standards outlined, should include information from a variety of sources.

Explain

Each group will present its solutions, including arguments from the evidence they gathered as well as their research, and make a recommendation to the audience (classmates at this point). Groups should prepare a written plan, and for their presentations, put their data and solutions on whiteboards (2 ft. × 3 ft. shower boards purchased at home improvement stores) or on paper posters, but with a "formal proposal" layout and quality. The teacher can facilitate the discussion (how the data were collected, what the internet research stated, what criteria might be used to evaluate the proposed solutions). The teacher should use key academic language listed earlier, having students incorporate the terms in their dialogue and presentations. Using whatever evaluation method works best in any given class, the group deemed to have developed the best presentation (including the best data collection procedure and design solution) should be given the opportunity to present to custodial staff and administration, along with copies of their written plan.

Elaborate

Encourage students to replicate this activity at home. Where financially feasible, have students try out their solutions (such as a water flow device). Encourage students to be alert to spotting water conservation opportunities in their neighborhoods and their community as a whole, such as broken sprinkler heads in yards, lawn sprinklers running too long and flooding the street, parks with standing water, and more.

Evaluate

Evaluation really takes place in all stages of a 5E instructional model. As you watch and listen to your students, you can discern their understandings of the purpose for the activity. Students should write a one-page (at minimum) explanation, independently, describing their research and proposed solutions to show completion of the activity's objectives.

Discussion and Argumentation

Students have the opportunity to support their proposed solutions using their research findings; audience members can be encouraged to ask for evidence when the groups present.

Differentiation

1. **Broader Access Activity**: Have students with lower language, ELA, and math skills work with students with stronger abilities. Students can be encouraged to create illustrations or other nonwritten representations of the process their group went through during the activity.

2. **Extension Activity:** Students can replicate the activity and develop a water conservation solution for their home, their parents' workplace, a local mall, or other venue.

3. **Modified Assessment:** Allow students to express their understanding orally, through artwork, music, videography, and more.

4. **Challenge Assessment:** Students could individually choose two of their peers' plans to evaluate, comparing and contrasting the strengths and weaknesses of the two plans.

Note: These are examples as illustrations of possible differentiation options. The actual adaptations you create will depend on the results of your preassessment and ongoing formative assessments of individual students.

CHAPTER 14

Activity 7: Only the Strong Survive!

Biology: Population Dynamics and Natural Selection

Every organism uses resources to grow and reproduce, and, every organism lives in relationship with the other living organisms and inanimate objects in their local area. This area is called an ecosystem.

In every ecosystem, there are limits to one or more of the resources that are required for an organism to live: limits to the availability of food, for example, or to the space available for shelter and security. As organisms consume food and other resources to survive, grow, and reproduce, other organisms compete for the limited resources.

Understanding the dynamics of this competition, and the effects of the competition on population growth and decline, is fundamental to understanding issues humans contend with as we interact within our Earth's ecosystems. This activity puts students in the roles of ecologists and wildlife biologists as they discover the principles of population dynamics and the role of genetic variation within a population of prey species.

Safety First

In every activity, we remind you to be certain that you understand the potential risks involved and are confident you can ensure your students' safety. Before attempting any of these activities in class, we recommend completing them yourself, and optimally with a teaching partner.

STEM

This activity capitalizes on the patterns in nature and models population dynamics and genetic variation within a species over time. As with all models, there are limitations and assumptions that allow the model to work. And, as with all models in the STEM disciplines, this model consistently demonstrates a major scientific concept that has been established through repeated observations across multiple areas and over a long period of years. Additionally, as with all models in STEM disciplines, this model will be continually improved upon over time.

This activity introduces students to a process in which they experience how scientific knowledge is developed and tested. Then, students use the scientific knowledge to predict and test additional questions and applications related to environmental quality, drinking water, and food sources. Opportunities to apply math standards emerge in the activity's use of statistics, tables, and graphing.

Tying It to *NGSS*

Analyze and interpret data to provide evidence for the effects of resource availability on organisms and populations of organisms in an ecosystem. (MS-LS2-1)

[Clarification Statement: Emphasis is on cause and effect relationships between resources and growth of individual organisms and the numbers of organisms in ecosystems during periods of abundant and scarce resources.]

Construct an explanation that predicts patterns of interactions among organisms across multiple ecosystems. (MS-LS2-2)

[Clarification Statement: Emphasis is on predicting consistent patterns of interactions in different ecosystems in terms of the relationships among and between organisms and abiotic components of ecosystems. Examples of types of interactions could include competitive, predatory, and mutually beneficial.]

Construct an argument supported by empirical evidence that changes to physical or biological components of an ecosystem affect populations. (MS-LS2-4)

[Clarification Statement: Emphasis is on recognizing patterns in data and making warranted inferences about changes in populations, and on evaluating empirical evidence supporting arguments about changes to ecosystems.]

Construct an explanation based on evidence that describes how genetic variations of traits in a population increase some individuals' probability of surviving and reproducing in a specific environment. (MS-LS4-4)

[Clarification Statement: Emphasis is on using simple probability statements and proportional reasoning to construct explanations.]

Use mathematical representations to support explanations of how natural selection may lead to increases and decreases of specific traits in populations over time. (MS-LS4-6)

[Clarification Statement: Emphasis is on using mathematical models, probability statements, and proportional reasoning to support explanations of trends in changes to populations over time.] [Assessment Boundary: Assessment does not include Hardy Weinberg calculations.]

Science and Engineering Practices

Analyzing and Interpreting Data

Analyze and interpret data to provide evidence for phenomena. (MS-LS2-1)

Constructing Explanations and Designing Solutions

Construct an explanation that includes qualitative or quantitative relationships between variables that predict phenomena. (MS-LS2-2)

Plan for safety by determining which engineering controls, safety procedures and personal protective equipment will be needed.

Engaging in Argument from Evidence

Construct an oral and written argument supported by empirical evidence and scientific reasoning to support or refute an explanation or a model for a phenomenon or a solution to a problem. (MS-LS2-4)

Evaluate competing design solutions based on jointly developed and agreed-upon design criteria. (MS-LS2-5)

Connections to Nature of Science

Science disciplines share common rules of obtaining and evaluating empirical evidence. (MS-LS2-4)

Analyzing and Interpreting Data

Analyze and interpret data to determine similarities and differences in findings. (MS-LS4-1)

Using Mathematics and Computational Thinking

Use mathematical representations to support scientific conclusions and design solutions. (MS-LS4-6)

Constructing Explanations and Designing Solutions

Apply scientific ideas to construct an explanation for real-world phenomena, examples, or events. (MS-LS4-2)

Construct an explanation that includes qualitative or quantitative relationships between variables that describe phenomena. (MS-LS4-4)

Connections to Nature of Science

Science knowledge is based upon logical and conceptual connections between evidence and explanations. (MS-LS4-1)

Disciplinary Core Ideas

LS2.A: Interdependent Relationships in Ecosystems

Organisms, and populations of organisms, are dependent on their environmental interactions both with other living things and with nonliving factors. (MS-LS2-1)

In any ecosystem, organisms and populations with similar requirements for food, water, oxygen, or other resources may compete with each other for limited resources, access to which consequently constrains their growth and reproduction. (MS-LS2-1)

Growth of organisms and population increases are limited by access to resources. (MS-LS2-1)

Similarly, predatory interactions may reduce the number of organisms or eliminate whole populations of organisms. Mutually beneficial interactions, in contrast, may become so interdependent that each organism requires the other for survival. Although the species involved in these competitive, predatory, and mutually beneficial interactions vary across ecosystems, the patterns of interactions of organisms with their environments, both living and nonliving, are shared. (MS-LS2-2)

LS2.C: Ecosystem Dynamics, Functioning, and Resilience

Ecosystems are dynamic in nature; their characteristics can vary over time. Disruptions to any physical or biological component of an ecosystem can lead to shifts in all its populations. (MS-LS2-4)

ETS1.B: Developing Possible Solutions

There are systematic processes for evaluating solutions with respect to how well they meet the criteria and constraints of a problem. (secondary to MS-LS2-5)

LS4.B: Natural Selection

Natural selection leads to the predominance of certain traits in a population, and the suppression of others. (MS-LS4-4)

LS4.C: Adaptation

Adaptation by natural selection acting over generations is one important process by which species change over time in response to changes in environmental conditions. Traits that support successful survival and reproduction in the new environment become more common; those that do not become less common. Thus, the distribution of traits in a population changes. (MS-LS4-6)

Crosscutting Concepts

Patterns

Patterns can be used to identify cause and effect relationships. (MS-LS2-2)

Graphs, charts, and images can be used to identify patterns in data. (MS-LS4-1), (MS-LS4-3)

Cause and Effect

Cause and effect relationships may be used to predict phenomena in natural or designed systems. (MS-LS2-1)

Stability and Change

Small changes in one part of a system might cause large changes in another part. (MS-LS2-4), (MS-LS2-5)

Connections to Nature of Science

Science assumes that objects and events in natural systems occur in consistent patterns that are understandable through measurement and observation. (MS-LS2-3)

Scientific knowledge can describe the consequences of actions but does not necessarily prescribe the decisions that society takes. (MS-LS2-5)

Tying It to *Common Core State Standards*, ELA

RST.6-8.7 Integrate quantitative or technical information expressed in words in a text with a version of that information expressed visually (e.g., in a flowchart, diagram, model, graph, or table). (MS-LS2-1)

RST.6-8.9 Compare and contrast the information gained from experiments, simulations, video, or multimedia sources with that gained from reading a text on the same topic. (MS-LS4-3), (MS-LS4-4)

WHST.6-8.1 Write arguments to support claims with clear reasons and relevant evidence. (MS-LS2-4)

WHST.6-8.2 Write informative/explanatory texts to examine a topic and convey ideas, concepts, and information through the selection, organization, and analysis of relevant content. (MS-LS2-2)

SL.8.1 Engage effectively in a range of collaborative discussions (one-on-one, in groups, and teacher-led) with diverse partners on grade 8 topics, texts, and issues, building on others' ideas and expressing their own clearly. (MS-LS2-2)

SL.8.4 Present claims and findings, emphasizing salient points in a focused, coherent manner with relevant evidence, sound valid reasoning, and well-chosen details; use appropriate eye contact, adequate volume, and clear pronunciation. (MS-LS2-2)

Tying It to *Common Core State Standards, Mathematics*

MP.4 Model with mathematics. (MS-LS2-5)

6.RP.A.3 Use ratio and rate reasoning to solve real-world and mathematical problems. (MS-LS2-5)

6.EE.C.9 Use variables to represent two quantities in a real-world problem that change in relationship to one another; write an equation to express one quantity, thought of as the dependent variable, in terms of the other quantity, thought of as the independent variable. Analyze the relationship between the dependent and independent variables using graphs and tables, and relate these to the equation. (MS-LS2-3)

6.SP.B.5 Summarize numerical data sets in relation to their context. (MS-LS2-2)

Common Misconceptions

One common misconception in this subject area is that natural selection is the same as evolution. Evolution is the word used to describe change over time, and it is considered a theory. Biologists agree that there are five distinct and different processes that result in evolution. While there are passionate debates regarding how and what is taught regarding biological evolution in schools, there is less debate over teaching processes such as natural selection and population dynamics.

Another common misconception is that individuals within a species may adapt to an environmental factor. Adaptations are the result of genetic variation and are passed on to future generations if the adaptation will help the individual with that variation survive and reproduce. Some of the common misconceptions are related to nonscientific beliefs and a lack of understanding the scientific principles and data regarding natural selection and population dynamics.

If you search the internet, you will find abundant websites that provide a range of explanations and representations regarding genetic variation and natural selection. However, caution should be exercised regarding evaluating the sources of the information. NSTA offers excellent resources on natural selection and the other principles associated with evolution. Obviously we don't have the space in this book to provide all the information you may desire regarding this topic. Therefore, we suggest seeking additional information from the NSTA website (*www.nsta.org*) and the NSTA Learning Center (*learningcenter.nsta.org*; search for "natural selection" and "evolution").

Once again, for more on helping students break through scientific misconceptions, consider the resources developed by Dr. Philip Sadler and others associated with the Harvard-Smithsonian Center for Astrophysics: see "Minds of Our Own" and other free online resources and video resources at *www.learner.org/resources/series26.html*.

In each activity, regardless of subject matter, we also urge readers to consider Page Keeley's series about identifying the prior knowledge your students bring to your classroom: see *Uncovering Student Ideas in Science* (*http://uncoveringstudentideas.org*).

Objectives

By the end of this activity, students will have demonstrated the ability to

- collect, analyze, and pool data;
- compare data and data analyses to look for patterns;
- role-play as predators in a model ecosystem and also as ecologists and wildlife biologists as they collect and analyze data and make claims based on evidence regarding the population dynamics across varieties of a prey species;
- analyze data and use data tables and graphs to describe patterns in nature (e.g., the variation in the number of individuals across multiple generations of a population in the model ecosystem); and
- implement procedural consistency and emulate research teams working collaboratively to process large sets of observations.

Academic Language

Trend, cycle, population, specie, prey, predator, population dynamics, carrying capacity, offspring, generation, natural distribution, biostatistics, forceps, model, variation, genetic factors, environmental factors, resources, limits

Focus Question(s) (Scientific Inquiry)

- How will a population of a prey specie vary across a variation and multiple generations?
- What living and nonliving factors appear to affect changes in the population dynamics in successive generations?
- How much variation is there among individuals of the same species?
- How do scientists and engineers collect large data sets?
- What advantages do large data sets provide scientists and engineers, compared with small data sets?
- When do you know you have enough data to have a high degree of confidence to make a claim based on evidence?

Framing the Design Problem(s) (Engineering Practice)

A sample problem for students might be "Design a plan to enhance an environment to optimize the survival of an endangered species or to optimize the survival of humans." (The variation among humans could be geographic location, potable water, or food for good nutrition, for example.)

Refer to Figure 4.3 "The Engineering Design Process" (p. 45) for more information about how to frame the design problem for this activity.

Teacher Background

This activity requires careful data collection and provides a rich opportunity to explore and elaborate on the limitations of models and simulations by having students compare their activity data to real data sets available on the internet.

We suggest teachers research various internet resources starting with the NSTA Learning Center's information on natural selection and evolution; there also are many decent YouTube and Khan Academy videos, as well as TEDed resources to review content knowledge associated with natural selection, population dynamics, and evolution.

Preparation and Management

This is an activity that can be set up with minimal cost, be stored in a small space, and last for years. There are computer simulations for this concept. However, we suggest this "hands-on" role-play as the entry point to the concept. After completing this activity, students will be better able to appreciate and understand the simulations that are available on the internet. One potential extension of this lab would be to have students analyze simulations for strengths and weaknesses compared to their "hands-on" role-play activity.

The Prey

The prey species is a paper dot (the punch out from a paper hole punch used for 3-hole punching paper for a 3-ring binder). If you have access to an electric punch, it will make the set up much easier.

You will want to use construction paper or simply various colors of copy paper. Pastel colors work best, or match the colors that are similar and that contrast well with the environment (piece of cloth). We suggest you use colors that are found and that contrast to colors in the environment (fabric square described in the following section on the "environment").

Having a variety of prey variations is important. The number of variations determines the complexity of the activity. At a minimum, we suggest five different colors; at a maximum you could use up to 10 colors, but the analysis becomes significantly more complicated.

You will want to use sandwich-size resealable bags to hold the "stock" for each color variation of the prey species.

The Predators

Students play the role of predators in this modeling game. Allowing students to use only a tool such as forceps or tweezers to pick up the prey makes the process work consistently. In the Differentiation section, we provide a list of increasingly complex variations among the predators.

The Environment

The teacher will need to get a set of multicolored fabrics in the same sizes. (The size of the cloth determines the size of the ecosystem. We recommend a minimum size of 50 cm × 50 cm, and a maximum size of 1 m × 1 m. The level of "busy-ness" in the pattern and size of the pattern elements also plays a role in this setup.

Prep Time

30 minutes to gather materials

Schedule time for students to use computers for internet access

Teaching Time

90–100 minutes, or two or three teaching periods

Materials

- Environment (1 piece of multicolored cloth to lay down on the tabletop)
- Prey organisms (1 resealable plastic bag per variation of prey; each bag needs a minimum of 100 pieces—if you use 5 variations, then you will have 5 bags, with each bag containing 100 paper punches of a single color.)
- Forceps or tweezers (one pair per student who is role-playing a predator)
- 1 stopwatch (or a wall clock with a second hand to time predation periods)
- Lap notebook (to record population of each prey variation per generation of play)
- Whiteboard, markers, and eraser for each group of three students (large chart paper and felt pens can be used in place of whiteboards)
- Safety glasses or goggles

5E Instructional Model

Engage

Show PowerPoint slides of places that profile the effects of human activities and diminishing habitat, and therefore the growing number of endangered and extinct species, and also the effects of proper management to save some species (for example the recovery of the bald eagle populations in North America).

Preassessment

Ask students to do the following:

1. Share what they've heard about endangered species and the recovery of some populations.
2. Describe what they know about invasive species and their effects on the health of the environment and any impact on human populations.

These two questions may be given as writing or discussion prompts to assess student prior knowledge about natural selection and population dynamics. Another suggestion is to have students develop individual explanations, share them, and ask other students to determine what seems correct and incorrect and what evidence supports their stances. The goal is to determine students' familiarity with the content as well as their ability to use evidence to support their positions (i.e., basic argumentation skill); this helps you shape the rest of the activity and subsequent lessons according to student ability and needs.

Explore

Table group teams are ideally four students each, but not any larger. Each team member has a role, and the roles should rotate with each "generation" of play. (When the hunting session is completed for a generation, all students participate in counting and establishing the appropriate number of each variety of species to be set up in the environment for the next hunting season.)

The roles include

- one student who is the time keeper and environmental steward (this student sets up the environment with the proper number of each prey species, while other players look away to keep them from seeing the arrangement prior to the beginning of the hunting session), and

- three students who are predators (these students look away while the time keeper/environmental steward sets up the environment for another hunting session).

Generation 1

1. All students work together to count out 20 of each variation of prey organisms.

2. When a set of 20 organisms for each fur coat color has been set in piles on the edge of the environment (the fabric square), students identify each role and who will play the roles in the first generation of play.

3. The predators will look away as the timekeeper/environmental steward mixes all the various colored punches and distributes the "prey" fairly evenly across the environment.

4. When the timekeeper is ready, the predators will pick up a set of forceps or tweezers, and the predators begin hunting for a specified time limit. This is predetermined by the team, and could be 20 or 30 seconds or a time limit that seems appropriate for the size of the environment and other conditions that

might exist in your classroom. Limit the time such that the predators cannot collect 100% of a particular prey species variation in the first hunting session.

5. After the timed hunting session, all hunting ceases. All team members work on collecting the data. They count the total number of each variation still alive in the environment. Then, they calculate that every pair of survivors will produce 3 offspring. So, if there were only 10 of a particular color, that is 5 pairs, and each pair will add 3 new offspring, so the team could count out the original 10 survivors plus 15 new offspring for a total of 25 organisms in the environment for that color variation.

6. The previous timekeeper/environmental steward becomes one of the predators, and one of the predators becomes the timekeeper/environmental steward.

7. Typically, you will want to have 5–7 generations of play to begin to see patterns emerge.

Explain

Each team of scientists should produce a whiteboard explanation of the process they completed across the generations. These should be shared with the class. The teacher facilitates this discussion, and each group defends its information using evidence from the simulation. (This is an excellent opportunity to implement formative assessment practices for clarifying and reteaching topics and information that may be needed to improve the accuracy of what students learned.)

Each group creates a written analysis of what happened in its simulation; teams will compare the results with other teams and environments (fabric squares with different colors and patterns).

Elaborate

Students extend their claims and evidence to look at issues of habitat management across the various environments. Students can design a plan for protecting the vulnerable species. One real-life example we've used is the discussion of deer hunting season and its purposes from an ecosystem and population management perspective.

Evaluate

Students will be evaluated on their ability to complete the series of generations and remain consistent with the model. Additionally, students will analyze the variations between team results based on the data and observations. How did the different teams' experiences vary, and why?

Discussion and Argumentation

Students should be given multiple opportunities to voice their claims and evidence, as well as the opportunity to refine their claims based on class-pooled evidence. This dialogue may be facilitated by the teacher and should focus on the nature of scientific ways of knowing (systematic review of data and the processes used to collect the data). We are not pursuing one correct answer here! Instead, we suggest the students stay focused on exploring and articulating what they "know" and the evidence they have. If the students need to go back to the activity to run additional tests, that should be encouraged, time permitting.

This activity again invites students to evaluate whether the more data we have, the better and to critically review these questions:

- Is there a point at which you have enough data to make a claim, and a diminishing return on the investment of time and energy to continue collecting data?
- If there is a point of diminishing returns, how do you know when you are at the point where you don't need to collect additional data?

If Time Allows

One of the many important opportunities provided in this activity is the chance to help students experience the analysis and redesign aspects found in the *NGSS* domain of science and engineering practices.

This could be an extensive project, so again, it's as time permits and needs to be carefully scaled for the students who pursue it, but it may provide an excellent opportunity for students to deepen their understanding about their original findings and plans.

Our suggestion is for students to design another model to demonstrate the same principles and to provide a detailed written explanation for their design.

If time allows, the students could be provided the opportunity to run the same series of trials with their model, collect data, and provide an analysis of any variation or similarities to the data collected in the activity.

Depending on the time you have available, and any cross-curricular connections you might design into this activity, you could invest 2–4 class sessions in this extension.

Differentiation

1. **Broader Access Activity:** Limit the number of prey organisms of each type of prey coat color. By limiting the number of individuals of each variation, fewer generations will be required to get to the point of the activity. Students who

are easily distracted would have their attention focused only on the relevant items necessary for a particular stage of the activity, and have a reasonable expectation of completing the counting and sorting task around the same time that other teams are ready to record their data on the class data chart.

2. **Extension Activity:** Expand the options and have students explore other covariates, such as the size of the environment used and the density of prey species, as well as the number of predators, and so on. Also, students could investigate the role of genetics and environment on variability of organisms in more complicated models.

3. **Modified Assessment:** Modify the assessment to focus on the verbal, written, or diagrammatic elaboration of the principles in this lesson.

4. **Challenge Assessment:** Allow students to present a comparison of the effects of genetics and environment on variability among individuals in a species.

Note: These are examples as illustrations of possible differentiation options. The actual adaptations you create will depend on the results of your preassessment and ongoing formative assessments of individual students.

Activity 8: Saving the World—One Ecosystem at a Time

Biology: Ecosystems and Biological Diversity

Maintaining biodiversity within a healthy ecosystem is critical. What goes on in an ecosystem that makes it function?

In this activity, teams of students get an opportunity to research an ecosystem and design a solution to maintain the health of ecosystem services. Teams will evaluate the merits and the constraints of each solution and present oral arguments defending their solutions.

Safety First

In every activity, we remind you to be certain that you understand the potential risks involved and are confident you can ensure your students' safety. Before attempting any of these activities in class, we recommend completing them yourself, and optimally with a teaching partner.

STEM

Designing a solution to a scientific problem is very much an engineering STEM activity. Students will use technology to find information to help them learn about the science of ecosystems and come up with a rank-order list of solutions.

Tying It to NGSS

Evaluate competing design solutions for maintaining biodiversity and ecosystem services. (MS-LS2-5)

[Clarification Statement: Examples of ecosystem services could include water purification, nutrient recycling, and prevention of soil erosion. Examples of design solution constraints could include scientific, economic, and social considerations.]

Science and Engineering Practices

Evaluate competing design solutions based on jointly developed and agreed-upon design criteria. (MS-LS2-5)

Disciplinary Core Ideas

Biodiversity describes the variety of species found in Earth's terrestrial and oceanic ecosystems. The completeness or integrity of an ecosystem's biodiversity is often used as a measure of its health. (MS-LS2-5)

LS4.D: Biodiversity and Humans

Changes in biodiversity can influence humans' resources, such as food, energy, and medicines, as well as ecosystem services that humans rely on—for example, water purification and recycling. (Secondary to MS-LS2-5)

ETS1.B: Developing Possible Solutions

There are systematic processes for evaluating solutions with respect to how well they meet the criteria and constraints of a problem. (Secondary to MS-LS2-5)

Crosscutting Concepts

Stability and Change

Small changes in one part of a system might cause large changes in another part. (MS-LS2-4), (MS-LS2-5)

Tying It to *Common Core State Standards*, ELA

RST.6-8.8 Distinguish among facts, reasoned judgment based on research findings and speculation in a text. (MS-LS2-5)

RI.8.8 Trace and evaluate the argument and specific claims in a text, assessing whether the reasoning is sound and the evidence is relevant and sufficient to support the claims. (MS-LS-4), (MS-LS2-5)

Tying It to *Common Core State Standards*, *Mathematics*

MP.4 Model with mathematics. (MS-LS2-5)

6.RP.A.3 Use ratio and rate reasoning to solve real-world and mathematical problems. (MS-LS2-5)

Misconceptions

Misconceptions about ecosystems are as vast as the varieties of ecosystems! "Plants get their food from soil;" "The top of the food chain has the most energy because energy accumulates moving up the chain;" "Carnivores are ferocious and herbivores are passive;" there are far too many examples to list here. Instead, we refer you to this website— *http://ecomisconceptions.binghamton.edu/intro.htm*—for an extensive list of misconceptions.

There are also many websites that provide a range of explanations and accurate representations regarding ecosystems.

In addition to the NSTA Learning Center that offers substantial background information on ecosystems and habitats, for more on helping students break through scientific misconceptions, consider the resources developed by Dr. Philip Sadler and others associated with the Harvard-Smithsonian Center for Astrophysics: see "Minds of Our Own" and other free online resources and video resources at *http://www.learner.org/resources/series26.html*.

In each activity, we recommend Page Keeley's series about identifying the prior knowledge your students bring to your classroom: see *Uncovering Student Ideas in Science* (*http:// uncoveringstudentideas.org*).

Objective

By the end of this activity, students will have demonstrated the ablty to research and then present an evidence-based argument proposing various solutions to maintaining biodiversity and equilibrium in an ecosystem to preserve the ecosystem services.

Note: The *NGSS* standard for this activity requires research and comparison of existing technologies. To add a hands-on, minds-on "warm-up," time permitting of course, we offer a companion activity in parallel here. However, while it will add relevance and increase understanding of the science and engineering practices in the main activity, it will not meet the standard by itself.

Companion Activity Objective

Students will be able to create, observe and report on an ecosystem of their choosing. Students might select an ecosystem in a jar covered with plastic, bacteria in a petri dish, a terrarium with a critter, and so on.

Academic Language

Ecosystem, diversity, water purification, nutrient recycling, soil erosion, food webs, vigor, organization, resilience, ecosystem services

Focus Questions (Scientific Inquiry)

- How can the merits and constraints in maintaining diversity and health in an ecosystem be ranked?
- Why is diversity important in an ecosystem?
- Why is it important to preserve ecosystems even if people are not directly impacted by their degradation or destruction?
- What do you think an "ecosystem service" is? Can you give examples?
- How can ecosystem services be preserved?
- How do ecosystem services vary between different ecosystems?
- What are the economic implications of preserving ecosystems?
- What are the social considerations in preserving ecosystems?

Framing the Design Problem (Engineering Practice)

Students can pursue the following problem for this activity: "For each ecosystem targeted, design a solution to preserve the ecosystem services."

Refer to Figure 4.3 "The Engineering Design Process" (p. 45) for more information about how to frame the design problem for this activity.

Teacher Background

Biodiversity is the connection of living things to each other. Plants and animals that exist together in a particular area are said to live in an ecosystem (short for *ecological system*). These plants and animals interact with one another and with the nonliving elements of the area, such as climate, water, and soil. Ecosystems can be as small as the space under a log or as vast as the entire forest. Ecosystems are generally described in terms of the "services" they provide. Visit the National Wildlife Federation (*www.nwf.org*) for more background information regarding provisioning and regulating the cultural and supporting services of ecosystems.

Companion Activity

For a good example of an ecosystem students can create, study, and then investigate protective solutions to apply, see this sample activity in which students construct a classroom coral reef biome: *www.bsu.edu/eft/ecosystems/p/activities.html*.

In this companion activity, students create and then interact in a coral reef environment they build in class to experience visiting a coral reef and learn about ecosystem interdependence (i.e., roles of the reef inhabitants as producer, consumer, predator or prey, parasite or host, and other relationships that are mutually beneficial or competitive). Students explore how the balance of a reef ecosystem can be disrupted by human activities, inviting them to consider solutions.

Preparation and Management

Locate visuals of various ecosystems, aquatic as well as terrestrial, on the internet. Put visuals on PowerPoint slides to project—this helps engage students and bring alive different types of ecosystems (many of which your students may not identify as ecosystems per se).

Prep Time

60 minutes to locate internet visuals

Teaching Time

Three 45-minute class periods

Materials

- Internet visuals of ecosystems
- Whiteboard and markers for each group
- Student notebooks
- Access to internet and online resources
- (For the companion activity: student-collected materials for ecosystems, such as jars, plastic wrap, rocks, terrarium, and so on

5E Instructional Model

Engage

Invite students, seated in table groups, to watch slides of various ecosystems, with no prior discussion from the teacher. As each slide is projected, have groups discuss what they think they are seeing. At the end of the slides, ask students what all the slides had in common. Access students' prior knowledge by having them use their notebooks to make a list of their ideas about what a healthy ecosystem looks like and how they know it is healthy.

Preassessment

Ask the students to answer these two questions:

- How might you measure the health of an ecosystem?
- In what ways can humans have a positive effect on an ecosystem?

These sorts of questions—whether presented as writing prompts, table group starters, or in another small group discussion format—will help you gauge student prior knowledge regarding ecosystems. You can also consider their ability to support their ideas with evidence and details to determine their level of argumentation skill as you plan the rest of the activity and subsequent lessons on this topic.

Explore

In a whole-group brainstorm, make a list of all ecosystems students can think of: forest, desert, grasslands, mountain, aquatic (and the many sub-ecosystems). Leave space between each ecosystem so that students can sign up to investigate that ecosystem.

Let students sign up for the system they want to investigate, with no more than three students in each group. Students should record their ecosystems and group members in their notebooks.

Next, students need to work with their group members to do the following:

1. Brainstorm and research all the ways in which their ecosystem maintains its health. Students should look at the rigor, organization, and resilience of their system as well as the ecosystem services.

2. Make charts in their notebooks in preparation for their solutions. Students should chart their own ecosystem, identifying strengths, weaknesses, solutions, and so on (see example chart).

3. Students fill in their charts as they research solutions. Remind them to use the academic language for this activity.

4. Students discuss and then rank their solutions, with "1" being the most important solution to maintain the ecosystem services.

5. Students then transfer their notebook work to a whiteboard and discuss their plans to share orally with the full class.

Example: Coral Reef Ecosystem

Solution for Maintaining Health	Merits	Constraints (scientific, economic, and social considerations)	Ranking and Reasoning (lowest number = most important to do first)
Spectrographic imager	Takes pictures of coral from airplane for data comparison; can assess quickly	Expensive	
Predator-prey monitoring	Balance needed for health of ecosystem	Difficult to control and introduce species	
Coastal zone management	Control amount of development and industrial run-off	Cost of regulating	
SCUBA divers and snorkelers	Ecosystem service: cultural	Need to control and educate about coral reef care when diving. Hard to regulate diver activities.	

Explain

Student groups present their ecosystems and whiteboards, one at a time, to the rest of the class. The teacher invites other students to ask questions about the group's rankings. Be sure to apply the academic language in context as students explain their solutions.

Elaborate

Each group takes their top-ranked idea from their chart and draws a "to scale" diagram depicting their idea.

Evaluate

Students should write their explanation of the process they went through to rank-order their solutions.

Discussion and Argumentation

When students present their ecosystems and whiteboards, the class can ask questions and challenge their research findings and solutions, requiring the presenters to cite their evidence and sources.

Differentiation

1. **Broader Access Activity:** Provide access to a variety of types and levels of research materials, the internet, current magazines, and so on. Students should be taught how to find research from reliable sources. As you scaffold student research experiences, lower-achieving students might choose very simple information, while high-achieving students might choose to explore peer-reviewed research articles. The goal is for lower-achieving students to ultimately access the higher-quality research publications.

2. **Extension Activity:** Students can identify and research a local ecosystem in their city or area and complete a chart similar to the one done in their group. Ideally, students could use the chart to educate peers, make a public service announcement, or present their findings to relevant school or community leaders.

3. **Modified Assessment:** Allow students to demonstrate their understanding in verbal explanations or labeled drawings rather than the written process statements.

4. **Challenge Assessment:** Students can add a concept map of their ecosystem and various solutions, drawing arrows to show their interconnectedness.

Note: These are examples as illustrations of possible differentiation options. The actual adaptations you create will depend on the results of your preassessment and ongoing formative assessments of individual students.

Activity 9: Thinking Like Scientists and Engineers

Science and Engineering Processes: Testing and Communication

All kids can be scientists and engineers—so let's invite them to show it! This activity enables students to make sense of previous research and development work and capitalize on scientific and engineering trials to improve on a classic design: the paper airplane.

As scientists and engineers, students get to analyze previous research, design improvements on previous projects, collect data, draw conclusions, and make inferences based on "claims supported by evidence."

After these skills are in place, students begin the process of replicating what others have done and then support or refute the results of other scientists' work. This activity lets middle schoolers experience good science and engineering practices by *doing* them.

Safety First

In every activity, we remind you to be certain that you understand the potential risks involved and are confident you can ensure your students' safety. Before attempting any of these activities in class, we recommend completing them yourself, and optimally with a teaching partner.

STEM

This activity capitalizes on the application of past experiences and scientific findings to support innovative and creative improvements in designing and building a better paper airplane. As with all good science activities, there are limitations and assumptions that may be hidden or as plain as the Sun. And, as with many projects in the STEM disciplines, "Thinking Like Scientists and Engineers" introduces and integrates scientific concepts, mathematical processes, and engineering practices.

In this activity, students experience how scientific knowledge is developed and tested by using engineering processes to design, build, and improve on the design of a paper airplane. Then, students use their scientific and mathematical knowledge to predict and test additional applications of their new learning.

Tying It to *NGSS*

Define the criteria and constraints of a design problem with sufficient precision to ensure a successful solution, taking into account relevant scientific principles and potential impacts on people and the natural environment that may limit possible solutions. (MS-ETS1-1)

Evaluate competing design solutions using a systematic process to determine how well they meet the criteria and constraints of the problem. (MS-ETS1-2)

Analyze data from tests to determine similarities and differences among several design solutions to identify the best characteristics of each that can be combined into a new solution to better meet the criteria for success. (MS-ETS1-3)

Develop a model to generate data for iterative testing and modification of a proposed object, tool, or process such that an optimal design can be achieved. (MS-ETS1-4)

Science and Engineering Practices

Asking Questions and Defining Problems

Define a design problem that can be solved through the development of an object, tool, process or system and includes multiple criteria and constraints, including scientific knowledge that may limit possible solutions. (MS-ETS1-1)

Developing and Using Models

Develop a model to generate data to test ideas about designed systems, including those representing inputs and outputs. (MS-ETS1-4)

Analyzing and Interpreting Data

Analyze and interpret data to determine similarities and differences in findings. (MS-ETS1-3)

Engaging in Argument From Evidence

Evaluate competing design solutions based on jointly developed and agreed-upon design criteria. (MS-ETS1-2)

Disciplinary Core Ideas

ETS1.A: Defining and Delimiting Engineering Problems

The more precisely a design task's criteria and constraints can be defined, the more likely it is that the designed solution will be successful. Specification of constraints includes consideration of scientific principles and other relevant knowledge that are likely to limit possible solutions. (MS-ETS1-1)

ETS1.B: Developing Possible Solutions

A solution needs to be tested, and then modified on the basis of the test results, in order to improve it. (MS-ETS1-4)

There are systematic processes for evaluating solutions with respect to how well they meet the criteria and constraints of a problem. (MS-ETS1-2), (MS-ETS1-3)

Sometimes parts of different solutions can be combined to create a solution that is better than any of its predecessors. (MS-ETS1-3)

Models of all kinds are important for testing solutions. (MS-ETS1-4)

ETS1.C: Optimizing the Design Solution

Although one design may not perform the best across all tests, identifying the characteristics of the design that performed the best in each test can provide useful information for the redesign process—that is, some of those characteristics may be incorporated into the new design. (MS-ETS1-3)

The iterative process of testing the most promising solutions and modifying what is proposed on the basis of the test results leads to greater refinement and ultimately to an optimal solution. (MS-ETS1-4)

Crosscutting Concepts

Influence of Science, Engineering, and Technology on Society and the Natural World

All human activity draws on natural resources and has both short and long-term consequences, positive as well as negative, for the health of people and the natural environment. (MS-ETS1-1)

The uses of technologies and limitations on their use are driven by individual or societal needs, desires, and values; by the findings of scientific research; and by differences in such factors as climate, natural resources, and economic conditions. (MS-ETS1-1)

Tying It to *Common Core State Standards*, ELA

RST.6-8.3 Follow precisely a multistep procedure when carrying out experiments, taking measurements, or performing technical tasks. (MS-PS3-3), (MS-PS3-4)

RST.6-8.7 Integrate quantitative or technical information expressed in words in a text with a version of that information expressed visually (e.g., in a flowchart, diagram, model, graph, or table). (MS-PS3-1)

RST.6-8.9 Compare and contrast the information gained from experiments, simulations, video, or multimedia sources with that gained from reading a text on the same topic. (MS-LS4-3), (MS-LS4-4)

RI.8.8 Trace and evaluate the argument and specific claims in a text, assessing whether the reasoning is sound and the evidence is relevant and sufficient to support the claims. (MS-LS-4), (MS-LS2-5)

WHST.6-8.1 Write arguments focused on discipline content. (MS-PS3-5)

WHST.6-8.2 Write informative/explanatory texts to examine a topic and convey ideas, concepts, and information through the selection, organization, and analysis of relevant content. (MS-LS2-2)

WHST.6-8.7 Conduct short research projects to answer a question (including a self-generated question), drawing on several sources and generating additional related, focused questions that allow for multiple avenues of exploration. (MS-PS3-3), (MS-PS3-4)

WHST.6-8.8 Gather relevant information from multiple print and digital sources, using search terms effectively; assess the credibility and accuracy of each source; and quote or paraphrase the data and conclusions of others while avoiding plagiarism and following a standard format for citation. (MS-PS1-3)

SL.8.1 Engage effectively in a range of collaborative discussions (one-on-one, in groups, and teacher-led) with diverse partners on grade 8 topics, texts, and issues, building on others' ideas and expressing their own clearly. (MS-LS2-2)

SL.8.4 Present claims and findings, emphasizing salient points in a focused, coherent manner with relevant evidence, sound valid reasoning, and well-chosen details; use appropriate eye contact, adequate volume, and clear pronunciation. (MS-LS2-2)

Tying It to *Common Core State Standards, Mathematics*

MP.2 Reason abstractly and quantitatively. (MS-PS3-1), (MS-PS3-4), (MS-PS3-5)

MP.4 Model with mathematics. (MS-LS2-5)

6.SP.B.4 Display numerical data in plots on a number line, including dot plots, histograms, and box plots. (MS-PS1-2)

6.SP.B.5 Summarize numerical data sets in relation to their context. (MS-LS2-2)

7.RP.A.2 Recognize and represent proportional relationships between quantities. (MS-PS3-1), (MS-PS3-5)

Common Misconceptions

There are many misconceptions about how scientists and engineers do their work. One common misconception is that they follow a single scientific method, step by step, in their investigations and designs. It is true that scientists communicate through reports about their work, and these reports are in a standard format. However, the reports that are written by scientists do not typically represent a linear sequential process for how they arrived at their conclusions. Indeed, creativity is as integral to scientific process as linear logic and reasoning. Students new to active science, though, may associate science with workbooks and memorizing vocabulary terms—the idea that science and scientists are one-dimensional and not particularly interesting. This is a fun misconception for you to dispel!

For more information about how scientists and engineers search (creatively) for questions and answers, please check the numerous resources and video clips available at the NSTA Learning Center. TEDed is another good resource.

To help your students break through scientific misconceptions, again in this activity we suggest you consider the resources developed by Dr. Philip Sadler and others associated with the Harvard-Smithsonian Center for Astrophysics: see "Minds of Our Own" and other free online resources and video resources at *www.learner.org/resources/series26.html*.

In each activity, regardless of subject matter, we suggest readers to consider Page Keeley's series about identifying the prior knowledge your students bring to your classroom: see *Uncovering Student Ideas in Science* (*http://uncoveringstudentideas.org*).

Objectives

By the end of this activity, students will have demonstrated the abilty to

- express familiarity with science and engineering practices;
- read, decode, and analyze the written graphs of other scientists and engineers;
- evaluate lab reports for reasonableness and logic by asking the following questions: Does the report make sense? Does it follow the logical pattern of the data from the experiment? Could I explain it to someone?;
- use previous scientific and research data to design, build, and test an improved paper airplane; and
- implement procedural consistency and emulate research teams working collaboratively to process large sets of observations.

Academic Language

Logic, metacognition, replicate, test, verify, data, claim, evidence, inference

Focus Question(s) (Scientific Inquiry)

- What does the data say?
- What can be proven and what can't? Do data "prove" anything? (Here the teacher is working toward the concept that scientific data can be used to refute or support, but not to "prove.")
- What is fact and what is fiction? (Or, What kinds of observations and information can you trust, and why?)

Framing the Design Problem(s) (Engineering Practice)

Here is the problem for students: "Design, build, test, and defend an improved paper airplane. Use evidence from your tests to support the claim that your paper airplane is an improvement over previous models, and/or present a detailed analysis for the 'failures' that would need to be addressed before a next attempt at a design, build, and test series."

Refer to Figure 4.3 "The Engineering Design Process" (p. 45) for more information about how to frame the design problem for this activity.

Plan for safety by determining which engineering controls, safety procedures, and personal protective equipment will be needed in this activity.

Teacher Background

This activity requires careful data collection. It provides a rich opportunity to explore and elaborate on the limitations of projects and engineering "improvements" by having students compare their activity data to real data sets available via the internet, as well as data available from other research teams in the class.

Basic to good science teaching is establishing an atmosphere of investigative fervor. Early in the year, and certainly in the buildup to this activity, let students know that they *are* scientists and engineers and that you plan to use and reuse their skills. Skills that will pay the highest dividends and that will be celebrated are those that allow students to think and act like scientists and engineers. Those science and engineering practices are as follows:

- Asking questions (for science) and defining problems (for engineering)
- Developing and using models
- Planning and carrying out investigations
- Analyzing and interpreting data
- Using mathematics, information and computer technology, and computational thinking
- Constructing explanations (for science) and designing solutions (for engineering)
- Engaging in argument from evidence
- Obtaining, evaluating, and communicating information

Providing examples of these practices to which students can relate will help build a context for activities such as "Thinking Like Scientists and Engineers."

Students should become familiar with the fact that research always starts with a question or puzzling observation, a problem, or a desire to make improvements on a past design, tool, or process. Additionally, students should become comfortable with asking questions, defending or justifying their answers, and being skeptical. (As we say in Chapter 3, "Ask questions and question answers!")

These practices are applicable to any topic you teach, and they foster metacognition. By providing your students with lab settings that maximize the use of the greatest number of practices, you are helping them become true scientists and engineers. Teach them to think, question, and continue to attempt new thoughtful ways to improve on past designs.

In terms of background information for this activity, once again we suggest teachers research various internet resources (including the NSTA Learning Center—see "Skills of a Scientist/Science Process Skills") to review content knowledge associated with experimentation and written reports on scientific experimentation.

Preparation and Management

This activity is "low-to-no cost" if you use scrap paper.

Prep Time

15 minutes. Make copies of previous lab reports written by students you have had in the past. If you're a new teacher, borrow lab reports from colleagues. Be sure to remove all student names from the reports. These reports should be of high quality. (In the Extend section, students are given reports with vague or inconclusive analyses to evaluate.) Be sure to save what your students create in labs this year to use with future classes in this activity.

Teaching Time (for the first phase of the 5E series in this activity)

Two 45-minute periods. You may wish to use one day to have students analyze past experimental data and the second day for their presentations. The second 5E phase will require an additional 2 or 3 class periods, as described on pages 193–194.

Materials

- One copy of a completed student lab report from a previous year (name removed) per group of two or three students (see sample on p. 189). You may want each group to have a different lab report to analyze.
- Two lab reports from a previous year (names removed): one done poorly, one done well
- Document camera

Note: See additional materials for second 5E cycle listed on page 193.

Sample Lab Report by Students: Building and Flying a Paper Airplane

Question
How can we make a paper airplane travel farther?

Hypothesis
If we add weight to our plane, then it will fly farther.

Plan
(Put on safety glasses or goggles!)

1. Fold paper into the shape of a plane.
2. Fly it without a paper clip.
3. Add one small paper clip to the bottom center of the plane.
4. Fly the plane multiple times, marking distances.

Test

Without paper clip

Trial 1	Trial 2	Trial 3	Trial 4	Trial 5	Average
82 ft.	50 ft.	93 ft.	hit ceiling	80 ft.	67.4 ft.

With a paper clip

Trial 1	Trial 2	Trial 3	Trial 4	Trial 5	Average
60 ft.	74 ft.	88 ft.	62 ft.	90 ft.	74.8 ft.

Conclusion

Adding the paper clip does increase the distance of the flight. We need more space to fly!

5E Learning Cycle Model

Engage

Show the two lab reports one at a time. Ask the students which lab report they would like to claim as theirs and why. Listen closely to comments as the students tell you what they think is effective or ineffective based on the sample reports. This is your opportunity to access their prior knowledge about what a good lab report should look like, and concerning their broader familiarity with scientific process and practice.

Preassessment

If you use the above situation as a prompt for students to write their responses individually before sharing, you will have good insight on the individual thinking and prior knowledge for each student in your class, from which you can adjust instruction and better tailor your implementation of the activity.

Explore

Distribute a copy of a completed lab report to each group. (See Sample Report on p. 189 and the rubric on p. 191.) Ask the group members to review the work as scientists and engineers would. They should ask themselves the following questions:

1. Am I able to interpret the information on this document?
2. Does it make sense?
3. Can I replicate what the previous science team has done?
4. Do I agree with their conclusion(s)? Why or why not?

Tell the students, "Be prepared to share orally with the whole class. You will be asked to tell the class what your report is about and what reactions different members of your group had to the report. All members of the group must participate in the presentation."

Rubric for Evaluating "Building and Flying a Paper Airplane"

(1 = not at all; 10 = completely)

1. Did the team doing the test follow a procedure?

   1   5   10

2. Did the test seem appropriate for the hypothesis?

   1   5   10

3. Did the test seem fair?

   1   5   10

4. Did the students conduct enough trials? Did they gather enough data?

   1   5   10

5. Did the conclusion seem logical?

   1   5   10

6. Overall, how would you rate this team on the ability to think like scientists?

   1   5   10

Explain

Allow time for students to share what their group discovered by looking at the lab reports. Rich discussion can emerge from the presentation of information. Students should be encouraged to ask questions, challenge results, or offer alternative interpretations. The teacher should contribute to the conversation after students have had an opportunity to share their thoughts first, even writing down their thoughts about what a good lab report looks like. (This is an excellent opportunity to clarify and reteach information to improve the accuracy of what your students learned.)

Extend

Give students the opportunity to apply their scientific thinking by developing several investigation options, preferably extensions of activities completed in class, and allow them to choose one and carry it out.

Give the groups a new lab report with a vague or inconclusive analysis of the lab. Have each group discuss the interpretation, then make a list of facts that will support or refute the interpretation. Set up a debate between groups to argue in support or refutation of the interpretation of the data, based on the information in the lab report.

Evaluate

As discussed previously, teachers can use copies of lab reports from previous years (with student names removed). First-year teachers can borrow some reports from a colleague or invent a few. Make sure that each member of the group has his or her own copy of the report. When students complete a lab report evaluation, have each group trade it with another group. (Each report is passed only once.) If the report is done clearly and correctly, the new group should be able to follow the report and concur with it or refute it. If it is not written clearly, members of the receiving team may ask questions in writing or orally to clarify any uncertainties. Sometimes several groups run into the same questions; in that case, the teacher can address the questions with the whole class. This is a great way to assess your teaching as well as student learning.

Another assessment option is to have each group present and defend their analyses of the report.

Taking the lessons learned from the sample lab reports, students now engage in internet research to look for design improvements that use a standard 8.5″ × 11″ piece of paper as the material for an improved paper airplane as inferred from the sample lab report.

This leads students into a "next interaction" of scientific inquiry and engineering design practices.

Prep Time for 5E Cycle 2

30 minutes to gather materials

Schedule time for students to use computers for internet access.

Teaching Time

Two or three 45-minute teaching periods

Materials

- Multiple sheets of standard 8.5″ × 11″ paper
- Whiteboard, markers, and eraser for each group of three students (large chart paper and felt pens can be used in place of whiteboards)
- Safety glasses or goggles

5E Instructional Model, Round 2

This next 5E cycle uses the entire first cycle that has just been completed as the Engage portion of the next cycle.

Engage

Review the claims and evidence based on the first 5E cycle in this activity.

Ask students to

- share what they have heard about and experienced with paper airplanes and scientific and engineering reports, and
- explain what they know about design processes.

Explore

Students use the internet and other resources to find other examples of paper airplanes and examine data regarding their "improvements." Once the students have determined the reliability of the resources, the students in each team will propose an improvement. The proposal should include a written justification for their design improvements, as well as a prediction based on reliable sources that these improvements will be seen when they build and test their prototypes for improved paper airplanes.

Then, teams of two students each should design, build, and test their improved paper airplane and support their claims regarding improvement with data from their tests.

Explain

Each team of scientists should produce a whiteboard set of claims and evidence regarding their predictions about the design, anticipated results, and actual results.

These should be shared with the class as the teacher facilitates the discussion. After each group has defended its information, the information is shared in comparison to the whole-class common claims and evidence, and then each group can consider and explain any variation between the class data set and the claims the team made based on its evidence.

Elaborate

Each group prepares a written analysis of what happened in their tests, offering explanations based on their evidence and the whole-class data set. Next, the groups should rotate around the lab to compare their results with those of other teams using different design improvements.

Evaluate

Students will be evaluated based on their ability to complete the series of 5E cycles, clearly articulate the lessons learned regarding science and engineering practices, and compare their predicted success and realized success. How is it that even though you might find a "best design" set of instructions, your team may still have been unable to realize its anticipated success with improved data or outcomes? What does this tell us about the real experiences of scientists and engineers?

Discussion and Argumentation

Students should be given multiple chances to voice their claims and evidence and refine their claims based on class-pooled evidence. This dialogue may be facilitated by the teacher and should focus on the nature of scientific ways of knowing (i.e., systematic review of data and the processes used to collect the data). The goal is not to arrive at one correct answer. Instead, we suggest the students stay focused on exploring and articulating what they "know" based on the specific evidence they've discovered. If the students need to go back to the activity to run additional tests, they should be encouraged to do so, time permitting.

Here's another opportunity to ask students to evaluate the notion that the more data we have, the better. Is this true? Is there a point where you have enough data to make a claim, and there is diminishing return on the investment of time and energy to continue collecting data? Relate this back to the theme of this activity: What indicates to scientists and engineers when "enough is enough?"

If Time Allows

This activity offers a good opportunity for analysis and redesign, as recommended in the *NGSS* domain of Science and Engineering Practices.

This can be accomplished in less time than the original activity required. Students already have a familiarity with the materials and processes. Taking an additional 45–90 minutes can lead to substantial learning through the analysis and redesign process.

We suggest that you review the findings of the class, and allow one or two days to pass to allow students to use out-of-school time to explore options regarding improving their (safe!) design and construction of a new and improved paper airplane.

You could also introduce additional variables/categories for evaluating the airplanes, such as time aloft, stunts, and distance. This will not cost any additional instructional time, and will allow students to create and innovate endlessly—like real scientists and engineers! Then, you can have a one or two 45 minute sessions for students to construct and analyze data from another cycle starting with *Explore*, and then moving into *Explain*, *Elaborate*, and *Evaluate*.

Differentiation

1. **Broader Access Activity:** Limit time and/or resources for the activity. For example, provide 3–5 specific plans for the teams to select, and move directly to the second 5E cycle. By limiting the number of options, students who are easily distracted would be better able to focus on relevant options; shortening the scope of the activity increases the likelihood that these students can complete the task in approximately the same time as the other teams.

2. **Extension Activity:** Expand the options and have students explore other covariates, such as additional pieces of paper or other materials such as plastic trash bags, straws, and/or sticks and how they impact the test results.

3. **Modified Assessment:** Modify the assessment to focus on the verbal, written, or diagrammatic elaboration of the principals in this lesson.

4. **Challenge Assessment:** Allow students to present a comparison of which modifications led to improved designs across the class groups, and to propose new prototypes that would take these designs to a new level of improved performance. They need to incorporate the data from the various groups in their argumentation.

Note: These are examples as illustrations of possible differentiation options. The actual adaptations you create will depend on the results of your preassessment and ongoing formative assessments of individual students.

Activity IO: Super Sleuths Saving Civilization

Earth Science: Weather and Climate

P redicting natural disasters is critical for the survival of plant and animal populations. This activity puts students into the role of scientists charged with analyzing data sets to identify patterns in order to forecast when and where a disaster might strike.

STEM

This activity is a great way to integrate STEM in a lesson. Students are applying their knowledge of the science of natural disasters (earthquakes, volcanic eruptions, tsunamis), by using technology to implement their design plan and analyze mathematical data sets so they can make a prediction.

Tying It to *NGSS*

Analyze and interpret data on natural hazards to forecast future catastrophic events and inform the development of technologies to mitigate their effects. (MS-ESS3-2)

[Clarification Statement: Emphasis is on how some natural hazards, such as volcanic eruptions and severe weather, are preceded by phenomena that allow for reliable predictions, but others, such as earthquakes, occur suddenly and with no notice, and thus are not yet predictable. Examples of natural hazards can be taken from interior processes (such as earthquakes and volcanic eruptions), surface processes (such as mass wasting and tsunamis), or severe weather events (such as hurricanes, tornadoes, and floods). Examples of data can include the locations, magnitudes, and frequencies of the natural hazards. Examples of technologies can be global (such as satellite systems to monitor hurricanes or forest fires) or local (such as building basements in tornado-prone regions or reservoirs to mitigate droughts).]

Science and Engineering Practices

Analyzing and Interpreting Data

Analyzing data in 6–8 builds on K–5 and progresses to extending quantitative analysis to investigations, distinguishing between correlation and causation, and basic statistical techniques of data and error analysis.

- Analyze and interpret data to determine similarities and differences in findings (MS-ESS3-2)

Disciplinary Core Ideas

ESS3.B: Natural Hazards

Mapping the history of natural hazards in a region, combined with an understanding of related geologic forces can help forecast the locations and likelihoods of future events. (MS-ESS3-2)

Crosscutting Concepts

Patterns

Graphs, charts, and images can be used to identify patterns in data. (MS-ESS3-2)

Tying It to *Common Core State Standards*, ELA

RST.6-8.1 Cite specific textual evidence to support analysis of science and technical texts. (MS-ESS3-1), (MS-ESS3-2), (MS-ESS3-4), (MS-ESS3-5)

RST.6-8.7 Integrate quantitative or technical information expressed in words in a text with a version of that information expressed visually (e.g., in a flowchart, diagram, model, graph, or table). (MS-ESS3-2)

Tying It to *Common Core State Standards*, *Mathematics*

MP.2 Reason abstractly and quantitatively. (MS-ESS3-2), (MS-ESS3-5)

6.EE.B.6 Use variables to represent numbers and write expressions when solving a real-world or mathematical problem; understand that a variable can represent an unknown number, or, depending on the purpose at hand, any number in a specified set. (MS-ESS3-1), (MS-ESS3-2), (MS-ESS3-3), (MS-ESS3-4), (MS-ESS3-5)

7.EE.B.4 Use variables to represent quantities in a real-world or mathematical problem, and construct simple equations and inequalities to solve problems by reasoning about the quantities. (MS-ESS3-1), (MS-ESS3-2), (MS-ESS3-3), (MS-ESS3-4), (MS-ESS3-5)

Misconceptions

This topic is another one with a whole array of misconceptions attributed to each natural disaster: "We'd be better off without the greenhouse effect" (not really); "monsoons are characterized by rainfall" (not always); and so on—we found the following website helpful to gain background about the various forms of natural disasters: *http://tech42. net/science55/meteorology1.html.*

Again, the free Science Object resource from NSTA Learning Center with topics addressing Earth science and weather will help you gain perspective on common misunderstandings (and perhaps confront your own misconceptions). You can find Science Objects at the NSTA Learning Center (*http://learningcenter.nsta.org*). Select "middle school" and then "free resources."

Once again, for more on helping students break through scientific misconceptions, consider the resources developed by Dr. Philip Sadler and others associated with the Harvard-Smithsonian Center for Astrophysics: see "Minds of Our Own" and other free online resources and video resources at *www.learner.org/resources/series26.html.*

In each activity, regardless of subject matter, we also encourage readers to consider Page Keeley's series about identifying the prior knowledge your students bring to your classroom: see *Uncovering Student Ideas in Science* (*http://uncoveringstudentideas.org*).

Objectives

By the end of this activity, students will have demonstrated the ability to

- work collaboratively in teams of three researchers;
- analyze data to look for trends and patterns;
- use technology to design and then write a plan to analyze, interpret, and communicate data sets indicating predictions of future natural hazards in a CRAFT format (explained on pp. 200–201).

Academic Language

Earthquake, volcanic eruption, tsunami, hurricane, tornado, flood, data set, forecast, analysis

Focus Question(s) (Scientific Inquiry)

- What technology is in place now to predict natural disasters?
- How does mapping the history of natural disasters help scientists prepare for future natural disaster events?
- What would be the ideal technology to warn people of natural disasters?

- How do scientists develop solutions to problems? How might this apply to natural disaster events?
- Who benefits most from early warning systems?
- What factors would a city council need to consider in adopting an engineering solution to the problem of finding and implementing an early warning system?
- We'll do internet research for this activity; how do you know when you have a reliable online resource?

Framing the Design Problem(s) (Engineering Practice)

A sample problem might be: "Design a plan to analyze and evaluate data sets to ensure their reliability; the solution will accurately predict future natural hazard events."

Refer to Figure 4.3 "The Engineering Design Process" (p. 45) for more information about how to frame the design problem for this activity.

Teacher Background

Good predictions and warning systems that alert us about natural hazards save lives and habitats. A warning system allows people to act quickly to reduce injuries and minimize the damage to habitat and prevent economic loss. Technological and scientific advances in the past few decades have greatly increased the likelihood that people will have advance warning of disasters. The federal government has several systems in place to help predict and prevent natural disasters. These systems include establishing and maintaining reliable communication with the public.

Part of the challenge of this assignment is for students to find out this information and to identify and investigate reliable sources to use to collect, analyze, and organize data. Resist the temptation to help teams find and organize this information, only giving clues as needed; some whole-group discussion about identifying reliable sources may be necessary and will emerge from the focus questions.

This activity uses a CRAFT format sometimes used in language arts classes. CRAFT stands for:

- **Context:** Where does this take place? City council meeting? Newscast?
- **Role:** Who are you? Scientist? Concerned citizen? President?
- **Audience:** To whom are you writing? Senator? A company? A city council?
- **Format:** What format are you using for your writing? Diary entry? Letter to the editor? A scientific journal?
- **Task:** What are you writing about? A concern? A problem? A solution?

CRAFT for This Activity

- **Context:** In an area known for natural disasters, the public is increasingly concerned about the dangers, and there is a cry for better predictions of impending disasters.
- **Role:** You are part of a small team of scientists that's been commissioned by the city's mayor to collect data on past natural disasters in order to predict future natural disasters.
- **Audience:** Mayor and city council
- **Format:** Share information in an official written report.
- **Task:** With a team of three peers, students will choose a natural disaster to investigate. Design a plan for collecting and analyzing data sets summarizing past disasters (online or provided by the teacher in hard copy) to identify patterns in order to forecast when and where a disaster might strike. Write an official report to be presented to the mayor and city council.

Students need to be taught how to identify and use credible sources (i.e., criteria for valid print resources; websites that end in .edu, .org, .gov, and so on).

Students should be allowed to design their own data collection plan and determine how to analyze the data sets. This is part of the process of this activity and is illustrated in the Engineering Design Process example shown in Figure 4.3 on p. 45. Students will be defending their plan in a whiteboard sharing setting before writing their report for the city council.

Preparation and Management

Schedule the time and place for student teams to have internet access for research. If computers are not available to the students, prepare real data sets and descriptions of municipalities impacted periodically or consistently by natural disasters.

Prep Time

15 minutes if students have easy internet access; 1–3 hours if not, depending on the level of detail and embellishment in the teacher-generated data sets.

Teaching Time

40–45 minutes to identify teams and collect data;

40–45 minutes to come up with plan to share data collection and do a whiteboard activity;

40–45 minutes for teams to write their reports for mayor and city council (and these can also be published in the classroom or, as time allows, presented to the class or other audiences.)

Materials

- PowerPoint slides of places affected by natural disasters (three each of the natural disasters listed above in Academic Language), showing damage to plant and/or animal life, including human habitats
- Student access to computers with online websites or information printed from websites and teacher-generated "creative" materials
- Whiteboard, markers, and eraser for each group of three students (Large chart paper and chart marker pens can be used in place of whiteboards.)
- Paper for reports

5E Instructional Model

Engage

Show PowerPoint slides of places that have experienced natural disasters.

Ask students to share

- what they have heard about/experienced with these natural disasters, and
- what they know about early warning systems for these events.

These two *Engage* questions work in table groups or other small-group discussion settings; they may also be given as writing prompts to assess what your students already know about natural disasters. You might have students develop individual responses, share with the class, and ask other students to evaluate what seems correct and incorrect and what evidence they have for the positions they're taking. (This will help you also gauge their argumentation skills.)

Reading their writings or observing their conversations will give you valuable insights prior to the *Exploration* phase of the activity, and may cause you to adjust instruction based on the varying knowledge levels you identify.

Explore

Tell students that they will work in a team of three to investigate a natural hazard using the CRAFT format. Present and explain the CRAFT model. Post a CRAFT poster in the room displaying the steps for future reference.

Allow student teams to choose a natural hazard to explore, having an even distribution of natural hazards throughout the groups as much as possible. Each team should come up with its own plan about how it will collect and analyze data from reliable sources on the internet and organize the information to prepare a written report.

Explain

Each team of scientists will produce a whiteboard explanation of the process they went through to collect and organize their data. These should be shared with the class. The teacher facilitates the discussion.

Elaborate

After each group has defended its information on the whiteboard, the information should be written up in an official report to be presented to the mayor and city council. This report should be published in the classroom and, time permitting, presented to classmates and others (i.e., create a mayor and city council—be creative!). Check the ELA standards listed on page 198 to ensure you are working within those parameters.

Evaluate

Students will be evaluated based on their ability to design a plan for collecting and analyzing data sets related to past natural disasters, their ability to make a prediction based on the data collected, and the report written for the mayor and city council.

Discussion and Argumentation

During the whiteboard sharing (what we like to call a "board meeting"), each group can share their evidence supporting the accuracy of their data collection and be ready to defend their process and sources.

Differentiation

1. **Broader Access Activity:** Pair struggling students with partners who demonstrate proficiency in the activity, and/or provide the group with a readily

researchable topic (i.e., assign the specific natural hazard and provide a list of possible research sources for struggling students). Consider assigning struggling students to be "vice mayor" or "science team member," or on a "concerned citizens committee" for the added support roles.

2. **Extension Activity:** Design or improve on the design of an emergency warning system.

3. **Modified Assessment:** Instead of a written report, allow students to analyze data sets and use bullet points, illustrations, or charts to report data.

4. **Challenge Assessment:** Investigate one of the careers related to this field and present the information from that point of view, or present the report in a skit or newscast format.

Note: These are examples as illustrations of possible differentiation options. The actual adaptations you create will depend on the results of your preassessment and ongoing formative assessments of individual students.

Where Do I Go From Here?

Resources for Good Science in Middle School

At this point, we've looked at the middle school learner, described good science, discussed its classroom implementation, and illustrated it with sample activities. Where can you look to learn more?

About Our Resource Collection

Resources for good science are abundant, and the examples we offer in this section by no means constitute an exhaustive list. These are some of the resources that we've found useful with a broad audience of science teachers, veterans and novices alike. We believe the very best resource available to teachers seeking to incorporate learner-centered strategies is a mentor in your own school community—someone who has had success with good science instruction in the middle grades—because just like your students, you can learn much faster by seeing and doing it than by reading about it!

On a cautionary note, teachers need to be aware of current copyright restrictions on print materials intended for use and distribution at school. Considerations such as brevity of material used, acknowledgment of copyright, and scope of distribution limit what can and can't be used legally by teachers with students. For updates and details concerning the educational fair-use regulations, teachers are urged to consult the Library of Congress Copyright Office online at *http://lcweb.loc.gov/copyright*.

And a final recommendation, or caution, as you peruse our list of resources: The most important information for science teachers at all levels, and particularly for early career teachers, is about teaching science *safely*.

The Standards

National Governors Association Center for Best Proactices and Council of Chief State School Officers (NGAC and CCSSO). 2010. *Common core state standards.* Washington, DC: NGAC and CCSSO.

National Research Council (NRC). 2011. *A Framework for K–12 Science Education: Practices, crosscutting concepts, and core ideas.* Washington, DC: National Academies Press.

NGSS Lead States. 2013. *Next Generation Science Standards: For states, by states.* Washington, DC: National Academies Press. *www.nextgenscience.org/next-generation-science-standards.*

Publications of the National Science Teachers Association

Assessing Science Learning: Perspectives From Research and Practice (2008), edited by Janet Coffey, Rowena Douglas, and Carole Stearns. NSTA Press. A helpful guide for delving into how and what it means to assess students for understanding.

Crossing Borders in Literacy and Science Instruction: Perspectives on Theory and Practice (2004), edited by E. Wendy Saul, International Reading Association, and NSTA Press. A useful resource for exploring the integrative relationships between science, reading, and writing.

Exemplary Science for Building Interest in STEM Careers (2012), edited by Robert E. Yager. NSTA Press. A very useful resource for guiding students toward STEM careers, and appreciating why this is such an important mandate for middle school teachers.

Integrating Engineering and Science in Your Classroom (2012), edited by Eric Brunsell. NSTA Press. Includes multiple essays covering a range of specific topics; Brunsell's own chapter is particularly helpful in understanding the stages of the engineering design process.

The Lingo of Learning: 88 Education Terms Every Science Teacher Should Know (2003), by Alan Colburn. NSTA Press. A great resource for reviewing terms we may have forgotten or never knew that relate to science instruction.

National Science Teachers Association (NSTA) Position Statement: Science Education for Middle Level Students. *www.nsta.org/about/positions/middlelevel.aspx*

Professional Learning Communities for Science Teaching: Lessons From Research and Practice (2009), edited by Susan Mundry and Katherine E. Stiles. NSTA Press. This resource will help teachers make productive use of the time they spend with their team in PLCs.

Questions, Claims, and Evidence: The Important Place of Argument in Children's Science Writing (2008), by Lori Norton-Meier, Brian Hand, Lynn Hockenberry, and Kim Wise. NSTA Press. This book is very helpful in joining science and the ELA requirements of working through claims and evidence.

NSTA Ready-Reference Guide to Safer Science, Vol. 2 (2012), by Kenneth Russell Roy. NSTA Press. Ken Roy is considered the NSTA expert on safety and offers essential guidance for science teachers at all levels in this book.

The NSTA Reader's Guide to A Framework for K–12 Science Education (2012) by Harold Pratt. This guide provides an excellent overview of the *Framework*.

Safety in the Science Classroom: www.nsta.org/pdfs/SafetyInTheScienceClassroom.pdf. In our opinion, you can't learn too much about science safety, and this is another great resource.

Technology in the Secondary Science Classroom (2008), edited by Randy Bell, Julie Gess-Newsome, and Julie Luft. NSTA Press. A helpful guide for integrating technology in our STEM classrooms.

The Teaching of Science: 21st-Century Perspectives (2010), by Rodger Bybee. NSTA Press. Helpful for teachers, administrators, and curriculum developers to appreciate what young citizens should know, value, and be able to do to prepare for life and careers in the 21st century.

Whole Class Inquiry: Creating Student-Centered Science Communities (2009), by Dennis Smithenry and Joan Gallagher-Bolos. NSTA Press. A provocative approach in which the whole class takes ownership of an inquiry investigation, with the teacher very much in the background.

Web-Based and Multimedia Resources

The Annenberg/Corporation for Public Broadcasting site (*www.learner.org*) has high-quality professional development resources for science and math, including copyright-free reproducibles.

Bottle Biology (*www.fasiplanto.org/bottle_biology*) is a reliable website with some "bottle biology" materials available free as Adobe Acrobat PDF documents.

Carolina Biological (*www.carolina.com*): This site offers useful online teacher resources, including no-cost lesson plans, advice and guidance, teaching strategies, the *Carolina Tips* journal, and more.

Closing the Achievement Gap (*www.teachstream. com*): This video series features noted diversity consultant Glenn Singleton and covers three topics: opening the conversation on race, moving beyond diversity into a greater understanding of race, and taking action to close the achievement gap. Singleton's presentations are among the most powerful we've attended and this video set captures many of his essential themes.

Discovery Education (*www.discoveryeducation.com/teachers*): Abundant high-quality science content, readings, and instructional ideas aligned with standards.

Edutopia (*www.edutopia.org*): A varied and informative site with topics ranging from assessment to instructional designs, with abundant samples and lesson ideas.

Exploratorium (*www.exploratorium.com/IFI/index.html*): The Exploratorium Institute for Inquiry offers outstanding workshops, programs, and online support for inquiry science.

Exploratorium Institute for Inquiry (2006). "Introduction to Formative Assessment Facilitator's Guide." (*www.exploratorium.edu/ifi/workshops/assessing*). Helpful overview of formative assessment and ready-to-use materials and guidance for teacher trainers.

Flinn Scientific (*www.flinnsci.com*): Flinn's comprehensive website includes its online catalog with teacher-tested science activities and materials. Free content on the site includes lesson plans and other instructional resources, the *Flinn Fax* quarterly newsletter, and Flinn's extensive, highly regarded science safety support information.

Intel Education's Design and Discovery Curriculum (*http://educate.intel.com/en/designdiscovery*): This site, from one of the nation's leading technology innovators, provides middle–school–appropriate curriculum design and project ideas that are ready to use, along with STEM-specific resources.

Khan Academy (*www.khanacademy.org*): The ubiquitous site for content tutorials, Khan Academy is continually adding to its resources across a range of specific topics relevant to doing good science.

NASA (*www.nasa.gov*): NASA's site offers comprehensive Web-based free resources for teachers. In the past, free professional development from NASA has been arranged for teachers and schools—contact NASA for details. NASA's Planet Quest Exoplanet Exploration (*http://planetquest/jpl.nasa.gov*) provides an engaging window into space exploration.

National STEM Video Game Challenge (*www.stemchallenge.org*): Competition is at the center of the site's aims, which may not be for every school community; but this is an engaging website that speaks to the potential of gaming to support STEM education and objectives.

The Northwest Regional Education Laboratory (NWREL) (*www.nwrel.org/msec/pub.html*) offers many valuable and current free teacher resources, teaching strategies, insights from practitioners, and relevant research on inquiry teaching.

NGSS@NSTA (*ngss.nsta.otg*): This is a convenient and comprehensive resource for information regarding the *Next Generation Science Standards*.

PBS (*www.pbs.org*): The PBS site includes access to high-quality, online interactive simulations, lesson plans, and many free resources for teachers. In particular, the STEM Education Resource Center (*www.pbs.org/teachers/stem*) presents a huge collection of interactive and informative STEM resources including lesson plans, videos, and more.

The Private Universe (1987) (Videotape produced by Annenberg/CPB; *www.learner.org* or 1-800-532-7637): Why don't students learn science? This provocative video documents and explores the problem through interviews with teachers, middle school students, and science-illiterate but eloquent Harvard graduates. An eye-opening and ageless resource, this video is still widely used in university courses and science professional development.

Science, Technology, Education and Mathematics (STEM) Education Coalition (*www. stemedcoaltion.org*): A consortium of heavy hitters in science education provides resources and research on STEM education. Particularly useful for "making the case for STEM" presentations.

Science Careers (*http://sciencecareers.sciencemag.org*) is a job-seeking resource teachers can use in real time to illustrate career opportunities in academia, industry, and government. The site is an extension of the journal *Science* and affiliated with the American Association for the Advancement of Science (AAAS). Please see more on this topic in Appendix A (p. 215).

TEDed: Lessons Worth Sharing (*http://ed.ted.com*): This website supports the adaption of TED Talks as an entry point for student learning with deep understanding. This open-source resource provides the tools and countless iterations for teachers to borrow and revise for future use.

TERC Science and Math Learning (*www.terc.edu*): TERC is a nonprofit educational research and development organization dedicated to improving math and science learning and teaching. This site includes high-quality curriculum, professional development and technology information, and software for teachers.

Print Resources

Practices

America's Lab Report (2006), produced by the National Research Council. National Academies Press. This book presents a review of the challenges and how to support increasing lab experiences for public school students.

Bottle Biology (1993, 2003) by Mrill Ingram, Kendall/Hunt Publishing. This book includes a wide variety of activities and how-to information for making and using science equipment out of plastic bottles and other recyclable materials. Full of ideas for Earth, life, and physical science applications. See also a related website: *www.fastplants.org/bottle_biology*

Oooperative Learning (2009), by Spencer Kagan. Kagan. An encyclopedic guide to the theory, methods, and lesson designs of cooperative learning. This book is densely packed with suggestions and useful, readily applicable strategies for increasing student collaboration and group productivity.

Cooperation in the Classroom (1993), by David W. Johnson, Roger T. Johnson, and Edythe J. Holubec. Interaction. This is a foundational work by the authors most closely associated with the contemporary development and spread of cooperative learning. It includes fundamental principles that are still useful almost 20 years after the book was first published.

Differentiated Science Inquiry (2011), by Doug Llewellyn. Thousand Oaks, CA: Corwin Press. A practical and user-friendly method for applying differentiation to learner-centered science classrooms.

Designing Effective Science Instruction: What Works in Science Classrooms (2009), by Anne Tweed. NSTA Press and McREL. A strong resource for teachers and teacher leaders responsible for supporting science education transformation at the school level.

Formative Assessment and Standards-Based Grading (2010), by Robert Marzano. Marzano Research Laboratory. This book presents reviews of recent updates and practices associated with implementing assessment for learning and instruction.

A Framework for K–12 Science Education: Practices, Crosscutting Concepts, and Core Ideas. Committee on a Conceptual Framework for New K–12 Science Education Standards (2012), produced by the National Research Council's Board on Science Education, Division of Behavioral and Social Sciences and Education. Washington, DC: National Academies Press.

Integrating Differentiation and Understanding by Design: Connecting Content and Kids (2006), by Carol Anne Tomlinson and Jay McTighe. Alexandria, VA: Association for Supervision and Curriculum Development. Addresses the fundamental issues of adjusting instruction to student needs and establishing developmentally appropriate instructional objectives that align with a lesson's or unit's outcomes. A very useful blend of the main contributions of these two important practitioners.

The Nation's Report Card: Science in Action: Hands-on and Interactive Computer Tasks From the 2009 Science Assessment (2012), produced by the National Center for Education Statistics. This is a review from the Institute for Education Sciences at the U.S. Department of Education.

Negotiating Science: The Critical Role of Argument in Student Inquiry (2009) by Brian Hand, Lori Norton-Meier, Jay Staker, and Jody Bintz. Heinemann. This book outlines the how and why we need to have students be able to defend their data.

Ready, Set, Science: Putting Research to Work in K–8 Science Classrooms (2008) by Sarah Michaels, Andrew W. Shouse, and Heidi A. Schweingruber. National Academies Press. In our opinion, this is a must-read for all science practitioners and policymakers.

Rockets: Educator Guide (2008), produced by NASA. Kennedy Space Center. A resource for teachers filled with physical science activities that build student learning from science and engineering practices.

STEM Lesson Essentials, Grades 3–8: Integrating Science, Technology, Engineering, and Mathematics (2013), by Jo Anne Vasquez, Cary Sneider, and Michael Comer. Heinemann. Presents clear and useful approaches to understanding what STEM is and presents an instructional model enabling teachers to deliver authentic STEM lessons across the grade levels.

Taking Science to School: Learning and Teaching Science in Grades K–8 (2007), National Research Council. National Academies Press. Predates the *Framework* and *NGSS*, but still a useful resource for good science teaching concepts and principles.

Where Great Teaching Begins: Planning for Student Learning and Thinking (2011), by Anne Reeves. Marzano Research Laboratory. This book presents lesson design from the perspective of planning to enhance cognitive engagement for all students.

Assessment

Assessing Science Learning: Perspectives From Research and Practice (2008), edited by Janet Coffey, Rowena Douglas, and Carole Stearns. NSTA Press. A helpful guide for delving into how and what it means to assess students for understanding.

Enhancing Inquiry through Formative Assessment (2003), by Wynne Harlen. San Francisco: Exploratorium. A useful resource for understanding the relationship between hands-on, minds-on instructional methods and formative assessment.

Uncovering Student Ideas in Science, presenting use of formative assessment probes featuring the work of Page Keeley and her co-authors. These excellent, teacher-freindly books are found in the NSTA Learning Center.

Instructional Strategies for New and Experienced Middle-Level Teachers

The First Days of School: How to Be an Effective Teacher (2009), by Harry K. Wong and Rosemary T. Wong. Harry K. Wong Publications. Among the most practical and widely used text resources for new and experienced teachers, this book covers every aspect of the crucial first days of class. In middle school especially, the book's emphasis on procedures is a key to success.

Meet Me in the Middle: Becoming an Accomplished Middle-Level Teacher (2001), by Rick Wormeli. Stenhouse. A readable, timeless guide to middle school instruction in all subjects, written by a master middle grades teacher and packed with creative suggestions.

Classroom Management

Conscious Discipline (2001), by Becky Bailey. Association for Supervision and Curriculum Development. Presents a systematic approach to helping students grow through their experience with classroom management procedures. Effective and very appropriate to middle-grades learners. An "oldie but a goodie."

Discipline With Dignity (1999), by Richard Curwin and Allen Mendler. Association for Supervision and Curriculum Development. This is another timeless resource in our opinion, presenting strategies that are respectful and still relevant to teachers working with preadolescents. Widely read and applied by middle school teachers around the country.

Cognitive Science

Brain Rules: 12 Principles for Surviving and Thriving at Work, Home, and School (2008), by John Medina. Pear Press. Dr. Medina provides a comprehensive analysis of the latest research on brain function and activities we can take to optimize learning and healthy living. One of the best parts of this book is the fact that the author pulls no punches on clarifying the level of rigor and scientific research supporting each of the 12 rules.

Developing Minds: A Resource Book for Teaching Thinking (2001, 3rd ed.), edited by Art Costa. Association for Supervision and Curriculum Development. The most authoritative and comprehensive review of "brain research" available today. Eighty-five articles cover an extensive range of individual topics that will shape (or reshape) how teachers perceive the connection between how they teach and how students learn. There are numerous "brain-based" resources that emerged in the last decade, but few are as well-documented scientifically and still relevant to classroom practice as Costa's collection.

Graphic Novel

BANG! The Universe Verse: Book 1 and *It's Alive! The Universe Verse: Book 2* (2009, 2011), written and illustrated by James Dunbar. CreateSpace. A unique series of rhyming comic books that explain the most fundamental concepts regarding the origin of the universe and life on Earth. The third and final book in this series, *Great Apes!* is coming soon and will address the origin of the human race. The series can supplement study of the nature of science. As with any resource, teacher review and discretion are advised, as some of the series' themes must be presented carefully in conservative school communities. *NGSS* topics embedded in the stories include

- Motion and stability: Forces and interactions,

- Energy,

- From molecules to oorganisms: Structures and processes,

- Earth and human activity,

- Matter and its interactions,

- Ecosystems: Interactions, energy, and dynamics,

- Biological evolution: Unity and diversity,

- Heredity: Inheritance and variation of traits, and

- Biological evolution: Unity and diversity.

Periodicals

National Middle School Association (NMSA). *Middle School Journal* and *Middle Ground* (*www.nmsa.org/services/curriculum.htm*): These journals present crucial topics in middle-level education across content areas and include other topics such as management, accountability, leadership, student programs, and more. Available by subscription or as an NMSA member benefit.

National Science Teachers Association (NSTA) member journals *Science Scope* and *Science and Children* (*www.nsta.org/journals*): *Science Scope* is NSTA's journal dedicated to education at the middle and junior high school level; we find the elementary journal, *Science and Children,* to be extremely useful also. NSTA's periodicals feature teacher-tested methods, resources, training, and networking essential to good science instruction.

Associations and Workshops

American Association for the Advancement of Science (*www.aaas.org*): AAAS is dedicated to the advancement of science, enhancing cooperative science projects, improving understanding of and funding for science, and contributing to science education. The association's Project 2061 produced *Benchmarks for Science Literacy* (1993), the nation's first comprehensive blueprint for sweeping science education reform, which is still widely used in school systems today. The AAAS NetLinks (*http://sciencenetlinks.com*) provide ready-to-use K–12 lesson plans and activities, and AAAS also publishes the well-known professional journal *Science.*

American Modeling Teachers Association (AMTA; *http://modelinginstruction.org*): This organization has formed to support the implementation of modeling instruction as developed under the leadership of Dr. David Hestenes (Emeritus Professor of Physics, Arizona State University), the tireless efforts of Dr. Jane Jackson, and key modeling leaders across the country. Lead teachers continue to extend the original physics modeling instruction practices to freshman courses in physics, chemistry, and biology. Some of the pioneers in the original development and implementation of Modeling Instruction have conducted professional development for upper middle level teachers.

Association for Supervision and Curriculum Development (ASCD; *www.ascd.org*): Among the premiere sources of high quality instructional research and innovation,

ASCD features a rich website and journal (*Educational Leadership)* for members, including many STEM and science-related resources.

Leadership and Assistance for Science Education Reform (LASER; *www.si.edu/nsrc* click on "More about the LASER Center"): The National Science Resources Center, the Smithsonian, the National Academies, and several corporate sponsors offer this intensive one-week institute at a number of regional locations and annually in Washington, DC. LASER trains teams of educators, community members, and business partners in the five areas of science reform: curriculum, assessment, developing community and administrative support, professional development, and materials support.

Museum of Science, Boston, Technology and Engineering Curriculum Review (*www.mos.org/tec*): Provides an extensive set of teacher reviews of STEM instructional materials.

National Middle Level Science Teachers Association (NMLSTA, *www.nmlsta.org*): An NSTA affiliate serving middle-level science educators with resources and training geared to middle grades teaching and learning.

National Middle School Association (NMSA; *www.nmsa.org*): NMSA is a central resource for professional development, publications, research, and networking for middle-grades teachers in all content areas.

National Science Education Leadership Association (NSELA; *www.nsela.org*): NSELA provides information to its members on a wide variety of topics (student learning, safety, curriculum, technology, professional development, assessment, inquiry, and science education reform).

National Science Resources Center (NSRC; *www.nsrconline.org*): NSRC is operated by the National Academies of Science, National Academy of Engineering, the Institute of Medicine, and the Smithsonian Institution. NSRC disseminates information about exemplary science teaching resources, develops and disseminates curriculum materials, and provides training (especially leadership and technical assistance) to promote hands-on science.

National Science Teachers Association (NSTA; *www.nsta.org*): NSTA is a professional organization for K–16 science educators offering high-quality benefits and resources to members and nonmembers.

The Last Word

It's hard to finish writing—or reading—a book like this and feel as if it's really done. That said, we must say that this book is really intended as a beginning. As the preceding resource collection suggests, there are countless opportunities for teachers to extend their skills in good science teaching.

We hope the book serves you as a reference and that you pursue some of the other opportunities outlined in this chapter, particularly the workshops provided by the Exploratorium, NSTA, NSRC, and AMTA. The wealth of learning achieved through interactions with exceptional science educators at such workshops—not to mention the connections made with science educators around the country—has been the most important catalyst in our growth as science educators and learners—that is, next to doing science!

And finally, for those of you just starting to experiment with active science in the middle grades, remember that the transition is most effective when it comprises subtle shifts rather than a revolution. There will be mistakes along the way, but as Thomas Watson, IBM's founder, was fond of saying, "If you want to increase your success rate, double your failure rate."

Good luck, and have fun!

Teaching About STEM Careers

The Call for STEM Career Counseling ... in Your *Classroom*

How many of us deliver a unit about STEM careers, or take time to help students understand how scientific literacy can be an asset in *many* professions and vocations? There are reasons we might not do so, at least in a thorough and intentionally structured unit, and chief among them is lack of time given the pressures of supporting language and math skills beyond the science curriculum.

It's true that, without any explicit effort on our part as teachers, students may be turned on by doing good science in middle school, and some may go on to pursue STEM-related college degrees and careers in science as a result; but as an intentional and deliberate goal of our work in classrooms, inspiring children to become scientists "seems to be difficult to accomplish in terms of evidence for success and effectiveness over time. It is rarely included in textbooks or curriculum frameworks with specific ideas for trying and information to use in activities. Further, relatively few educators are able to discuss what they specifically do to meet the goal" (Yager 2012, vii).

In our experience, amidst the harried pace driven by the compounding demands of teaching science (and math and literacy skills) in today's outcomes-oriented education environment, very few middle school science educators take time to *teach* our students about STEM careers. Some schools and districts still organize career fairs that may include scientists and engineers, or parents who work in technology fields, and we know of a few such events that even focus on STEM careers.

But when it comes to engaging in activities in the classroom, leveraging your relationship with your students, and investing precious instructional time to inspire and influence them about a STEM future—this just isn't happening very much. Yet.

To be fair, we're pretty sure there's little district- or site-level training (or expectation) to do so. Middle school science teachers don't often view themselves as career counselors, and that's unfortunate, because research has demonstrated that subject teachers are the most useful source of information for children to consult about careers, and interest in pursuing a science career is formed by age 14 (Osborne et al., 2012).

The *Framework* suggests that with an emphasis on STEM education, some workforce development will happen passively: "Although the primary rationale for including engineering practices is not to recruit more engineers, the explicit inclusion of engineering and technology opens the door to curriculum materials that communicate to

APPENDIX A

students the broad spectrum of career opportunities that includes not only scientists but also technicians, engineers, and other careers that require knowledge and abilities in the STEM fields" (NGSS Public Release II, January 2013, p. 2).

We also know that traditionally American science education has had a stifling effect on student aspirations to pursue STEM careers. "Indeed, research would suggest that the standard approach to science education is only successful at engaging somewhere between a fifth and a third of high school students with the further study of science (Osborne et al. 2012, p. 52).

If we are to prepare a STEM workforce, we must understand what a STEM workforce entails. The U.S. Commerce Department's Economic and Statistics Administration (ESA) developed this outline of STEM jobs:

- Professional and technical support occupations in the fields of computer science and mathematics, engineering, and life and physical sciences
- Computer and information systems managers, engineering managers, and natural sciences managers
- Science occupations, including science technicians; engineering and surveying occupations, including engineering technicians and drafters; computer occupations ranging from computer support specialists to computer software engineers (Langdon, et al. 2011)

The ESA list of professions includes 50 occupation codes divided into four categories: computer and math, engineering, physical and life sciences, and STEM managerial occupations. See Table A.1 for the ESA's compilation of STEM professions.

Beyond the partial listing in Table A.1, note that other conceivably STEM-related professions (such as teaching!) fell outside the ESA's umbrella.

Science Careers (*http://sciencecareers.sciencemag.org*) is a job-seeking resource teachers can use in real time to illustrate career opportunities in academia, industry, and government. The site is an extension of the journal *Science*, and affiliated with the American Association for the Advancement of Science (AAAS).

The argument for both the need and influence of STEM preparation emerges from compelling ESA data about STEM careers and college majors:

- In 2010, there were 7.6 million STEM workers in the United States, representing about 1 in 18 workers.
- STEM occupations are projected to grow by 17.0% from 2008 to 2018, compared to 9.8% growth for non-STEM occupations.

- STEM workers command higher wages, earning 26 percent more than their non-STEM counterparts.
- More than two-thirds of STEM workers have at least a college degree, compared to less than one-third of non-STEM workers.
- STEM degree holders enjoy higher earnings, regardless of whether they work in STEM or non-STEM occupations (Langdon et al., July 2011).

Table A.I. A Partial Listing of STEM Occupations

Computer and Math Occupations	Physical and Life Sciences Occupations
Computer scientists	Agricultural and food scientists
Computer programmers and systems analysts	Biological scientists
Computer software engineers	Conservation scientists and foresters
Computers support specialists	Medical scientists
Database administrators	Astronomers and physicists
Network and computer systems administrators	Atmospheric and space scientists
Network systems and data communications analysts	Chemists and materials scientists
Mathematicians	Environmental scientists and geoscientists
Operations research analysts	Physical scientists
Statisticians	Agricultural and food science technicians
Miscellaneous mathematical science occupations	Biological technicians
	Chemical technicians
Engineering and Surveying Occupations	Geological and petroleum technicians
Surveyors, cartographers, and photogrammetrists	Nuclear technicians
Aerospace engineers	Other life, physical, and social science technicians
Agricultural engineers	
Biomedical engineers	**STEM Managerial Occupations**
Chemical engineers	Computer and information systems managers
Civil engineers	Natural sciences managers
Computer hardware engineers	Engineering managers
Electrical and electronic engineers	
Environmental engineers	
Industrial engineers, including health and safety	
Marine engineers and naval architects	
Materials engineers	
Mechanical engineers	
Mining and geological engineers, including mining safety engineers	
Nuclear engineers	
Petroleum engineers	
Engineers, all other	
Drafters	
Engineering technicians, except drafters	
Surveying and mapping technicians	
Sales engineers	

Source: Langdon et al. 2011.

Table A.2. STEM Undergraduate Degree Fields and Majors

Computer and information systems	**Engineering Majors**
Computer programming and data processing	Environmental engineering
Mathematics	Geological and geophysical engineering
Applied mathematics	Industrial and manufacturing engineering
General engineering	Materials engineering and materials science
Aerospace engineering	Mechanical engineering
Biological engineering	Metallurgical engineering
Architectural engineering	Mining and mineral engineering
Biomedical engineering	Naval architecture and marine engineering
Chemical engineering	Nuclear engineering
Civil engineering	Petroleum engineering
Computer engineering	Miscellaneous engineering
Electrical engineering	Engineering technologies
Engineering mechanics	Engineering and industrial management
Animal sciences	Electrical engineering technology
Food science	Industrial production technologies
Plant science and agronomy	Mechanical engineering–related technologies
Soil science	Miscellaneous engineering technologies
Environmental science	Military technologies
Biology	
Biochemical sciences	**Physical and Life Sciences Majors**
Botany	Genetics
Molecular biology	Microbiology
Ecology	Pharmacology
	Physiology
Technology Majors	Zoology
Computer science	Miscellaneous biology
Information sciences	Nutrition sciences
Computer administration management and security	Neuroscience
Computer networking and telecommunications	Cognitive science and biopsychology
	Physical sciences
Math Majors	Astronomy and astrophysics
Statistics and decision science	Atmospheric sciences and meteorology
Mathematics and computer science	Chemistry
	Geology and Earth science
	Geosciences
	Oceanography
	Physics
	Nuclear, industrial radiology, and biological technologies

Source: Langdon et al., 2011)

Beyond the ESA's list of "orthodox" STEM occupations in Table A.1 (and the sampling of STEM college fields and majors in Table A.2), there are untold support, clerical, and administrative careers that fortify, connect, and complete the STEM career listing, and indirectly involve or rely on some degree of STEM education. As Jonathan Gerlach reminds us,

> We cannot forget the larger segment of industry that relies on STEM. Construction, transportation, and even the hospitality industry rely on a STEM-developed workforce. Whether it's understanding how an engine works, or plotting trucking routes, the advanced level of technical knowledge and problem-solving capability needed for these positions have become obstacles that did not exist 10 years ago. This explains why industries view career and technical education as a key piece of STEM education. Students must be prepared for any path they choose in life, whether it is directly into a STEM career or studying a specialized STEM field in college. (2012)

Table A.3. Average Hourly Earnings of Full-Time Private Wage and Salary Workers in STEM Occupations by Educational Attainment, 2010

	Average hourly earnings		Difference	
	STEM	Non-STEM	Dollars	Percentage
High school diploma or less	$24.82	$15.55	$9.27	59.6%
Some college or associate degree	$26.63	$19.02	$7.61	40.0%
Bachelor's degree only	$35.81	$28.27	$7.54	26.7%
Graduate degree	$40.69	$36.22	$4.47	12.3%

Source: Beede and Langdon. 2011.

One glaring flaw in capturing any list of STEM-related professions, even a partial one like Table A.1, is recognizing that STEM jobs are morphing and growing in parallel with the astonishing changes in the STEM disciplines. The 21st-century challenge of preparing children for jobs that haven't been invented came alive for us in researching this part of the book—if we'd encountered a job listing for "Cloud Computing Administrator" or "Social Media Manager" when we wrote the first edition of this book in 2003, we'd have been clueless about what these professionals do for a living! But while we recognize that STEM's explosive growth will add exponentially to the careers we're referencing here, it only serves to reiterate that the skills shaped by good science are broadly applicable and relevant to the futures of children we're teaching now.

Truly, we could expand the reach of STEM career education to span across the professions and jobs that will rest on the so-called 21st-century skills—teamwork, problem-solving, creativity, resilience, curiosity, and so on—because these are the skills nourished and demanded by good science.

Ultimately, STEM learning outcomes transcend workforce development. As the *Framework* proclaims,

> Understanding science and engineering, now more than ever, is essential for every American citizen. Science, engineering, and the technologies they influence permeate every aspect of modern life. Indeed, some knowledge of science and engineering is required to engage with the major public policy issues of today as well as to make informed everyday decisions, such as selecting among alternative medical treatments or determining how to invest public funds for water supply options. (2012, p. 7)

To support this aim, the spectrum of skills and habits of mind that you're building by teaching good science is summarized in Figure A.1 below.

Figure A.I. STEM Learning Outcomes

- Active collaboration
- Presentation skills
- Literacy
- Content skills and knowledge
- Research skills
- Work ethic and effort
- Critical thinking and problem solving
- Social and cross-cultural skills
- Technology skills
- Writing skills

Source: Wojnowski, Charles, and Warnock. 2012, p. 71.

How do middle school science teachers promote and inform children about STEM careers? One effective strategy, short of a full-blown career fair, is to provide students direct experience with individuals involved in STEM-related jobs. This can involve inviting parents and community members to your class to present in individual or panel question-and-answer sessions, or having STEM professionals view and critique student projects or showcased work products as judges and VIPs. Silicon Valley's tech firms encourage their employees to do community service and outreach in schools—in part to promote science education—and this practice is increasingly common across science and engineering industries nationwide.

Teachers will find willing partners awaiting in private sector companies eager to contribute their time and talent to support STEM workforce development. We've seen career-shadowing field trips, guest speaker and teacher relationships, and even grant funding for programs and equipment emerge from these corporate partnerships. It takes some initiative on your part, but the possibilities (and the rewards for your students) are robust.

For more on encouraging students to pursue STEM careers, see NSTA's *Exemplary Science for Building Interest in STEM Careers* (Yager 2012).

References

American Management Association. 2010. Critical Skills Survey. *www.amanet.org/news/AMA-2010-critcal-skills-survey.aspx*.

Beede, D. N., and D. S. Langdon. 2011. *Understanding and expanding the STEM workforce*. Washington, DC: U.S. Department of Commerce Economics and Statistics Administration, Office of the Chief Economist.

Brunsell, E., ed. 2012. *Integrating engineering and science in your classroom*. Arlington, VA: NSTA Press.

Gerlach, J. 2012. STEM: Defying a simple definition. *NSTA Reports. www.nsta.org/publications/news/story.aspx?id=59305*.

Koenig, K., and M. Hanson. 2012. Fueling interest in science: An after school program model that works, in *Integrating engineering and science in your classroom,* ed. E. Brunsell, 183–188. Arlington, VA: NSTA Press.

Langdon, D., et al. 2011. STEM: Good jobs now and for the future. ESA Issue Brief #03-11. U.S. Department of Commerce Economics and Statistics Administration.

National Research Council (NRC). 2012. *A framework for k–12 science education: Practices, crosscutting concepts, and core ideas.* Committee on a Conceptual Framework for New K–12 Science Education Standards. Board on Science Education, Division of Behavioral and Social Sciences and Education. Washington, DC: The National Academies Press.

NGSS Public Release II, January 2013. Appendix I: Engineering Design, Technology, and the Applications of Science in the Next Generation Science Standards.

Osborne, J., et al. 2012. Educating students about careers in science: Why it matters. In *Exemplary science for building interest in STEM careers,* ed. E. Yager, 51–62. Arlington, VA: NSTA Press.

APPENDIX A

Wojnowski, B., K. Charles, and T. Warnock. 2012. Why STEM? Why now? The challenge for U.S. education to promote STEM careers. In *Exemplary science for building interest in STEM careers*, ed. E. Yager, 63–80. Arlington, VA: NSTA Press.

Yager, R., ed. 2012. *Exemplary science for building interest in STEM careers.* Arlington, VA: NSTA Press.

Glossary of Good Science Terms

5E instructional method. Engage, explore, explain, elaborate, and evaluate; a framework for structuring good science lessons.

argumentation. The process of supporting a scientific claim using evidence. Students learn how to engage in argumentation as part of their path to thinking like scientists and engineers.

authentic assessment. The use of assessment strategies (beyond paper-and-pencil tests) that involve demonstration of learning—for example, projects, portfolios, and presentations. Authentic assessment is an integral component of good science and middle school instruction.

constructivism. A learning theory centered on the notion that instruction is most effective and lasting when students are actively involved in "constructing" meaning, assimilating new information by building on prior knowledge and experience, rather than passively memorizing material that may or may not be connected to any other knowledge.

cooperative learning. As with "real" science, good science instruction is most often collaborative; collaborative student groups are most effective when they operate with clear procedures, well-defined expectations for the outcome of their work, and active monitoring by the teacher.

differentiation. Good science instruction is differentiated according to the needs and ability levels of students, which can vary significantly among students in a single middle school classroom due to the wide range of cognitive, social, physical, and emotional development of adolescents. Differentiation directs a group of students toward the same outcomes; however, the amount and nature of support the teacher provides is individualized and is determined based on preassessment and formative assessment data.

discrepant event. An event that is contrary to what is expected to happen; an anomaly; often the catalyst for an investigation, raising the question "Why did this occur?" and leading to more questions.

engineering design. A scientific process in which students define problems in terms of criteria and constraints, research the problem to build awareness and understanding, develop and test possible solutions, and refine their solutions through a redesign phase.

extensions. Teachers of good science are encouraged to challenge their students to apply and extend their investigations, making connections to their everyday lives and pursuing questions of their own in further teacher-approved investigations.

focus questions. Teachers use focus questions to initiate and guide the progress of investigations, with the teacher serving as a coinvestigator.

formative assessment. Ongoing assessment, which includes observation and authentic assessment strategies in addition to pencil-and-paper tests, used to help teachers gauge student understanding as well as the effectiveness of lessons and activities. Formative assessment must be used to improve instruction for it to be meaningful.

inquiry. An approach to scientific investigation that begins with a question, problem, or observation and is followed by hypotheses testing or systematic observation and data collection and the reporting of findings. Often, one inquiry leads to others as learning leads to new questions. Teachers can assume more or less control of an inquiry activity according to difficulty, safety, time, or other concerns.

integration. Good science actively incorporates instruction and practice in the essential skills of mathematics and the language arts. When taught effectively, science is the "crossroads of curriculum."

kits. Also called "units," kits contain materials necessary for students to conduct active scientific investigations and may be developed and built by teachers for specific customized units; commercially prepared kits are available prepackaged from a number of vendors (see Chapter 18).

metacognition. The awareness of one's learning or thinking processes; teachers help students develop this awareness.

science lab notebooks. Integral to good science, notebooks enable students to record lab observations, practice journaling skills, reflect critically; also, they serve teachers as a means of formative assessment.

scientific method. A cyclical process that may be entered at any point. We shy away from defining a "single" scientific method, but the process is systematic and often includes observations, hypotheses, problems, experimental designs, data collection and analyses, conclusions, questions for further investigations, and revisions.

STEM. An acronym for science, technology, engineering, and math; describes an integrated approach to science teaching and learning in which at least two of the STEM disciplines are engaged in an activity. Sometimes also called STEAM or STEM-D to acknowledge the important role played by art and design in the engineering process. Note that "technology" refers broadly to human-made systems used to solve problems, extending beyond merely digital devices and electronic media.

subtle shifts. A phrase coined by San Francisco's Exploratorium to describe the successful, gradual, low-stress transition from traditional teacher- and text-centered science instruction to activity-based teaching—teachers are advised to move at a reasonable pace and not attempt (or expect) a revolution in their classrooms.

Sample Lab Report Form

Name _____ Date _____

Problem:

_____

_____

Hypothesis:

_____

_____

_____

_____

_____

_____

Experimental Design:

_____

_____

_____

_____

_____

_____

_____

_____

Variables:

_____

_____

_____

Results:

Data Chart:

Graph:

Analysis:

_____

_____

_____

_____

_____

Conclusion:

_____

_____

_____

_____

_____

Sample Design Challenge Report Form

Name _____ Date _____

Design Challenge Considerations:

A. Problem: (What is the challenge?)

_____

_____

_____

B. Constraints (limits) on the design (size, materials, mass, etc.):

_____

_____

_____

Design Options: (What are some possible solutions to the problem?)

1. _____

_____

2. _____

_____

3. _____

Which design did you select to build and test? Give at least one reason you believe your design has the best chance of winning the challenge.

_____

_____

_____

Results of the test of your design:

Data Chart:

Graph:

Analysis (including 2 or 3 proposed revisions that can improve the results, based on evidence from your tests):

_____

_____

_____

_____

What is your revised design plan (redesign)? What reasons support the revised design and make you confident the new design will be successful? (Include specific evidence from your first test and results and/or the results of other designs that were tested.)

_____

_____

_____

Redesign Test Results:

Data Chart:

Graph:

Analysis #2 (including proposed additional revisions to improve results, based on evidence from the second round of testing):

_____
_____
_____
_____

Conclusion:

_____
_____
_____
_____

Science Lab Safety Rules

1. Safety Acknowledgement form must be signed and returned before students are allowed to participate in labs.

2. Students must dress properly when engaged in lab activities—long hair must be tied back, no loose sleeves or jewelry, wearing closed toe and heeled shoes. Students must wear lab aprons as instructed.

3. Students must conduct themselves in a responsible manner at all times. Horseplay, practical jokes, and pranks have no place in the science classroom.

4. All written and verbal instruction must be followed. Students should ask their teacher if they do not understand the instructions.

5. Students must know what to do in case of an emergency in the science classroom, including knowing where safety equipment is and how to use it and the location of emergency exits

6. Equipment should not be touched until students are told to do so, and it should be treated properly.

7. No eating, drinking, chewing gum, or tasting anything in the science classroom.

8. Students should wash their hands before and after a lab.

9. Safety goggles should be worn when instructed (especially when using chemical or projectile hazards).

10. Lab work areas should be neat and free of unnecessary materials.

11. All chemicals used in the science room are to be considered dangerous. There should be no smelling or touching chemicals unless instructed by the teacher.

12. Science equipment or materials should never be taken out of the science classroom by students without teacher permission.

13. Students should never enter a science storage closet at any time.

14. Glassware should be handled with care; broken glass should never be touched by students, but reported to the teacher and disposed of properly.

15. Accidents must be reported immediately to the teacher; students should stand clear of any accident until it is resolved.

16. Extreme caution must be taken when a flame, burner, or hot plate is involved. Do not leave these items unattended.

17. At the end of the lab, work areas and equipment should be cleaned and waste disposed of as instructed.

18. A hard copy of Safety Data Sheets should be visible in the classroom and on file in the nurse's office.

For additional information, see *www.nsta.org/portals/safety.aspx*.

Index

Index

Index

Index